LEGACIES AND CHANGE IN POLAR SCIENCES

Global Interdisciplinary Studies Series

Series Editor: Sai Felicia Krishna-Hensel
Interdisciplinary Global Studies Research Initiative,
Center for Business and Economic Development,
Auburn University, Montgomery, USA

The Global Interdisciplinary Studies Series reflects a recognition that globalization is leading to fundamental changes in the world order, creating new imperatives and requiring new ways of understanding the international system. It is increasingly clear that the next century will be characterized by issues that transcend national and cultural boundaries, shaped by competitive forces and features of economic globalization yet to be fully evaluated and understood. Comparative and comprehensive in concept, this series explores the relationship between transnational and regional issues through the lens of widely applicable interdisciplinary methodologies and analytic models. The series consists of innovative monographs and collections of essays representing the best of contemporary research, designed to transcend disciplinary boundaries in seeking to better understand a globalizing world.

Also in the series

Global Cooperation
Challenges and Opportunities in the Twenty-First Century
Edited by Sai Felicia Krishna-Hensel
ISBN 978-0-7546-4678-5

The New Security Environment
The Impact on Russia, Central and Eastern Europe
Edited by Roger E. Kanet
ISBN 978-0-7546-4330-2

Sovereignty and the Global Community
The Quest for Order in the International System
Edited by Howard M. Hensel
ISBN 978-0-7546-4199-5

International Order in a Globalizing World
Edited by Yannis A. Stivachtis
ISBN 978-0-7546-4930-4

Legacies and Change in Polar Sciences
Historical, Legal and Political Reflections on The International Polar Year

Edited by

JESSICA M. SHADIAN
High North Center for Business and Governance,
Bodø Graduate School for Business, Norway

MONICA TENNBERG
University of Lapland, Finland

Routledge
Taylor & Francis Group

LONDON AND NEW YORK

First published 2009 by Ashgate Publishing

Published 2016 by Routledge
2 Park Square, Milton Park, Abingdon, Oxon OX14 4RN
605 Third Avenue, New York, NY 10017

First issued in paperback 2020

Routledge is an imprint of the Taylor & Francis Group, an informa business

British Library Cataloguing in Publication Data
Legacies and change in polar sciences : historical, legal
  and political reflections on the International Polar Year.
  -- (Global interdisciplinary studies series)
  1. International Polar Year, 2007-2008. 2. Polar
  regions--Research--International cooperation.
  I. Series II. Shadian, Jessica Michelle. III. Tennberg,
  Monica.
  333.7'072011-dc22

Library of Congress Cataloging-in-Publication Data
Shadian, Jessica Michelle.
  Legacies and change in polar sciences : historical, legal and political reflections on the
International Polar Year / by Jessica M. Shadian and Monica Tennberg.
     p. cm.
  Includes bibliographical references and index.
  ISBN 978-0-7546-7399-6 -- ISBN 978-0-7546-9174-7 (ebook) 1.
International Polar Year, 2007-2008. 2. Polar regions--Research--International
cooperation I. Tennberg, Monica. II. Title.
  G587.S44 2009
  508.311072--dc22
                                                                    2009004329

ISBN 13: 978-0-367-74016-0 (pbk)
ISBN 13: 978-0-7546-7399-6 (hbk)

Contents

List of Figures

Notes on Contributors

Dr. Michael Bravo is senior lecturer at the University of Cambridge and a Fellow of Downing College. He is based at the Scott Polar Research Institute, Department of Geography, where he is head of the Circumpolar History and Public Policy Research Group. Bravo has an interdisciplinary background with a humanities Ph.D. (Cantab 1992) in the history and philosophy of science, building on a technical background with a B.Eng. (Carleton 1985) in satellite communications engineering. Bravo has written extensively on the role of scientific research in the exploration and development of the Arctic, exploring issues in the philosophy of experiment such as the nature of precision and calibration. In his co-edited book *Narrating the Arctic* (2002), he explored the implications of the Arctic's extraordinary historical diversity through the lens of the Scandinavian Arctic. Bravo is currently leading an International Polar Year project making a comparative study of the uses of polar research stations. Under this umbrella his team has begun to assemble the first overview of the creation of polar research stations from the 1820s to the present day, linking them to other crucial developments in science such as the laboratory revolution and the invention of international scientific years (e.g. IPYs). His current concerns include the recent rise of cryo-politics, the ethics of environmental regulation, and the dangers that sea ice loss pose to the political rights and traditions of the Arctic's inhabitants.

Dr. Sanjay Chaturvedi is currently the coordinator, Centre for the Study of Geopolitics, Department of Political Science and honorary director, Centre for the Study of Mid-West and Central Asia, at Panjab University, Chandigarh. His area of specialization is the theory and practice of geopolitics; with special reference to Polar regions, Indian Ocean and South Asia. He was awarded the Nehru Centenary British Fellowship, followed by a highly coveted Leverhulme Trust Research Grant, to pursue his post-doctoral research at Scott Polar Research Institute, University of Cambridge, England, from 1992 to 1995. Dr. Chaturvedi has been a recipient of several visiting fellowships abroad, including Department of Geography at University of Durham, UK, Columbia University Institute for Scholars, Reid Hall, and Maison des Sciences de l'Homme, Paris (under International Programme of Advanced Studies); Faculty of Law, University of Sydney, Australia; Ben Gurion University of the Negev, Israel (under Distinguished Visitors Programme); and Henry L. Stimson Centre, Washington DC, USA. He was also awarded a research grant by the Australia-India Council, in connection with his project, 'Australian-Indian Perspectives on Antarctic and Ocean Governance: Interactions, Linkages and Opportunities'. Chaturvedi also serves on the steering committee of

International Geographical Union (IGU) Commission on Political Geography for the term 2004–08.

Dr. Marcus Haward is an Associate Professor in the School of Government and Institute of Antarctic and Southern Ocean Studies at the University of Tasmania. Haward serves as the programme leader, policy program, Antarctic Climate and Ecosystems Cooperative Research Centre (ACE CRC), University of Tasmania. He has also held visiting appointments at the Australian National University, Australian Maritime College and at Dalhousie University, Halifax, Canada and has published widely in the areas of Antarctica, fisheries management, coastal and oceans governance and public policy. Dr. Haward has been a member of Australian delegations to the Antarctic Treaty consultative meeting, CCAMLR and to APEC Fisheries and Marine Resources Conservation working group meetings and is on the editorial board of the international journal *Ocean and Coastal Management.*

Dr. Rob Huebert is an Associate Professor in the Department of Political Science at the University of Calgary. He is also the associate director of the Centre for Military and Strategic Studies. Dr. Huebert has also taught at Memorial University, Dalhousie University, and the University of Manitoba. His areas of research interest include: international relations, strategic studies, the law of the sea, maritime affairs, Canadian foreign and defence policy and circumpolar relations. He publishes on the issue of Canadian Arctic security, maritime security, and Canadian defence. His work has appeared in *International Journal*; *Canadian Foreign Policy*; *Isuma: Canadian Journal of Policy Research* and the *Canadian Military Journal*. He was also a co-author of the *Report To Secure a Nation: Canadian Defence and Security into the 21st Century*; and co-editor of *Commercial Satellite Imagery and United Nations Peacekeeping* and *Breaking Ice: Canadian Integrated Ocean Management in the Canadian North*. He also comments on Canadian security and Arctic issues in both the Canadian and international media.

Dr. Julia Jabour teaches Antarctic law and policy in the Institute of Antarctic and Southern Ocean Studies' Undergraduate and Honours programmes. She is also deputy leader of the Policy Program at the Antarctic Climate and Ecosystems CRC and leader of the project 'Management of Marine Living Resources in the Southern Ocean'. She has the roles of graduate research coordinator (about 60 M.Sc. and Ph.D. candidates) and honours coordinator (about 15 candidates). Dr. Jabour was a member of the Australian delegation to Antarctic Treaty Consultative Meetings in St. Petersburg in 2001 and Madrid in 2003. She has been to Antarctica five times, visiting Casey Station in the Australian Antarctic Territory, the Antarctic Peninsula and the Ross Sea region of East Antarctica. She has lectured on Antarctic cruise ships departing from Ushuaia in Argentina, Hobart and Lyttleton in New Zealand. Her research interests include Antarctic, international and maritime law and policy; international relations; tourism; science communication; and ethics.

Dr. Annika E. Nilsson has a background in biology and journalism and has worked professionally as a science writer since the 1980s. She has written extensively, including reporting and popular science books on climate science and policy as well as other topics with close science–society links. As a freelance journalist, she also has experience in the science–policy interface in the Arctic, where she has written popular science summaries of scientific assessments of pollution issues and edited an assessment of human development. She has an interdisciplinary Ph.D. from Linköping University. Sweden, with a dissertation about science and policy in the Arctic Climate Impact Assessment (ACIA). She is currently a research fellow at Stockholm Environment Institute where her research is about learning at the science-policy interface.

Dr. Donald R. Rothwell has been Professor of International Law at the ANU College of Law, Australian National University since July 2006 and previously taught at the University of Sydney where he held the post of Challis Professor of International Law. His research has a specific focus on law of the sea, law of the polar regions, use of force, and implementation of international law within Australia. Rothwell has been a member of expert groups for UNEP, UNDP, IUCN and the Australian government, and in 2006 chaired the Report of the Sydney Panel of Independent International Legal Experts on Japan's Special Permit ('Scientific') Whaling Under International Law.

Dr. Jessica M. Shadian is a senior research fellow at the High North Center for Business and Governance at the Bodø Graduate School for Business in Bodø, Norway. Prior to this Shadian completed a post doctoral fellowship at the Barents Institute in Arctic Norway. Shadian holds a Ph.D. in political science and international relations (global governance) from the University of Delaware, United States, and has been a visiting researcher at the Arctic Centre in Rovaniemi, Finland; Department of History of Technology, Royal Institute for Technology, Stockholm, Sweden; CIERA (Centre interuniversitaire d'études et de recherches autochtones), Laval Université, Quebec, Canada; and the Scott Polar Research Institute, Cambridge University (November 2004 to August 2005), where she completed her dissertation research investigating the implications of indigenous autonomy on Western conceptions of sovereignty and Arctic governance as manifest in the work of the Inuit Circumpolar Conference. Her current research projects include: member of accepted IPY project – *History of Polar Field Stations* based at Cambridge University in the UK; *Arctic Norden: Science, Diplomacy and the Formation of a Post-War European North* based at KTH; and a member of IPY project – *Large Scale historical exploitation of Polar Areas (LASHIPA)*.

Dr. Monica Tennberg is a Research Professor at the Arctic Centre, University of Lapland Finland. Her background is in political science, international relations. She defended a doctoral dissertation in 1998 about the development of states-indigenous peoples relations during the establishment of the Arctic Council. The dissertation was based on Michel Foucault's ideas about knowledge and power and their application to the theories of international regimes in international environmental cooperation. She has continued to study international environmental cooperation in the Arctic, especially lately Arctic climate politics and performance of international environmental cooperation in Northwest Russia. Dr. Tennberg and her research team participate in a pan-Arctic IPY project on community adaptation and vulnerability to climate change in Arctic regions (CAVIAR) with funding from Finnish Academy (2007–09).

Dr. Consuelo León Woppke is the Director of the Hemispheric and Polar Research Center, Fundación Valle Hermoso, Viña del Mar, Chile. She holds an M.A. and a Ph.D. in Diplomatic History from the Southern Illinois University at Carbondale. Dr. León Wöppke was a Fulbright scholar and an American Association of University Women (AAUW) doctoral fellow. She was a founding member and, later, the president of the Chilean Association for North American Studies (ACHEN). Dr. León Wöppke has published extensively on Chilean Antarctic policy, US–Chilean–Argentine Antarctic relations, and the concept of the Western Hemisphere. Her current, government-sponsored research project addresses U.S.–Chilean relations in the period of crisis from 1927–31. She has recently authored a book dealing with the national, international, and hemispheric contexts of Chilean Antarctic policy, in addition to editing two others dealing with Chilean press coverage of the Antarctic dispute.

Dr. Urban Wråkberg (M.Sc., Ph.D.) is the Research Director of the Barents Institute, Kirkenes, Norway. Wråkberg has 20 years' experience of research in the social sciences and the humanities on the subarctic, Arctic and Antarctica. He was among the initiators of the Swedish Programme for Social Science Research in the Polar Regions in 1995 and scientific leader of its multidisciplinary expeditions and fieldwork in Svalbard in 1997, 1998, 1999 and 2000, based on logistics provided by the Swedish Polar Research Secretariat. In 2002 he undertook field investigations in Northeast Greenland and in 2003 in Tierra del Fuego, Argentina. Wråkberg returned to Svalbard in 2005 as expert adviser on historical archaeology of the Finnish IPY Kinnvika recognisance expedition. He has published in Swedish, Norwegian and English on the social construction of knowledge of the polar regions since 1990, including *The Centennial of S.A. Andrée's North Pole Expedition* (1999), *An Arctic Passage to the Far East* (2002) and *Antarctic Challenges* (2004) as well as the opening chapter of volume one of the three-volume multi-authored *Norsk Polarhistorie* (2005) (published in English as *Into the Ice: The History of Norway and the Polar Regions*).

Preface
Legacies of Polar Science

Michael Bravo

Today the polar regions are attracting many new interested readers. They are once again in the public eye to a degree not seen for decades. The causes of the interest and even sensationalism surrounding the poles comes from the intersection of three global phenomena: the impact of climate change on the Arctic, the volatility of international commodity prices in response to unprecedented demand for minerals and hydrocarbons, and the political contest for control of polar resources. The environmental and security challenges that these present have rapidly become the focus of discussion and debate in academic and policy circles.

The rules and regimes by which the polar regions are governed has become a matter of intense political maneuvering as states and other groups seek to advance their respective interests. In both polar regions, political actors are jockeying to position themselves for opportunities and dangers that accompany the exploitation of natural resources and the concomitant environmental regulation. Today more than ever, politically interested parties in the polar regions represent themselves as enlightened conservationists seeking to protect the globe's natural heritage. Yet the language of conservation is masking a wide range of political interests that are not always easy to decode. Researchers in both the natural and social sciences are being invited to join the new game of polar geopolitics to help devise strategies to redefine the polar regions to these ends. It is ironic that at precisely the moment when there is an international collective effort to harness environmental research in the Arctic under the guise of disinterested knowledge, there appears to be less transparency than ever about the use of academic research in the social and natural sciences by policy makers. As academic researchers we have an ethical challenge to preserve the intellectual neutrality and critical objectivity of our universities.

Until recently it seemed self-evident to many researchers in the natural sciences and policy making, that the polar regions, particularly the Antarctic, could be conceptualized as a laboratory, as though they existed in natural isolation from the rest of the world. But the metaphor of the laboratory, insofar as it implies a bounded environment where relationships between specific variables can be controlled by controlling or excluding others, has always been contingent on political and economic circumstances. While the laboratory metaphor continues to enjoy some currency, it is clear that the wider significance of scientific practice in the polar regions is coming under the spotlight to an unprecedented degree. This attention is closely linked to powerful interests in the energy sector, state interests in energy security, and a range of organizations anxious about the impacts of climate change and other environmental destruction. Although questions of sovereignty and conflict are probably being overplayed, the response of markets to potential energy reserves has unsettled assumptions about regional governance and the prerogative of the eight members of the Arctic Council to control the science–policy interface with reverberations for governance in Antarctica. This in turn has created an opportunity to explore the assumptions that underwrite political authority in the polar regions and as well as their capacity to control the commissioning and deployment of experts which would seem to account for a large slice of the economic pies in the two polar regions.

Arctic ecosystems, though fragile, are believed to be ecologically amongst the most resilient on the planet because of their capacity to shift to new states in the face of extreme environmental pressures. Nevertheless they are seen to be in a highly precarious situation today because of the impacts of temperature increases, sea ice melting, permafrost melting and toxic contamination. These environmental pressures are global in scale and originate largely outside the Arctic, and as such, they constitute the precise conditions in which scientific research is being used to justify a wide range of political interventions, some neo-liberal and others protectionist, in the name of global environmental stewardship. Through the multiple political uses of expert evidence, strategies and models for the political mobilization of science, whether in the guise of seismic and hydrographical surveys of continental shelves, resource development impact benefit agreements, or the adoption of marine protection strategies, are being translated and deployed between regions on a global scale.

The human, social and physical sciences today are more important than ever for what they tell us about the changing living conditions of our planet. Many aspects of the mechanisms linking biological, ocean and atmospheric systems remain poorly understood by scientists, although long-term monitoring by experienced observers and autonomous instruments may help to define crucial new research problems at these interfaces. Nevertheless, scientific research is producing a fascinating picture that reveals the extent to which the Arctic is connected to other regions through global-scale carbon and heat exchange systems through the oceans, atmosphere and human economic activity. These resource, carbon and heat exchange cycles cut across traditional ways of classifying and organizing knowledge systems (e.g. social, economic, political and physical). This is leading to the realization that the political and policy synergies between the polar regions are far less substantial than assumed by previous generations of polar explorers and scientists.[1] As the comparative importance of the connections between Arctic ecosystems and the temperate regions become better understood, historians looking back at the present day may describe the Arctic as entering a new 'post-polar' era, and the notion of 'polar science' as an artifact of the twentieth century, when in the light of the

---

1   The similarities between the polar regions tend to relate to significant areas of research in geophysical field sciences such as geomagnetism, auroral studies, and glaciology, rather than ecosystems, human settlement, cultural history, or politics.

alliances between geopolitics and the field sciences, it made sense to group the Arctic and Antarctic as two of a kind.

The challenges of environmental regulation and conservation in the polar regions in the decades ahead will require fresh thinking about the organization of research methodologies. It is unlikely that the same approaches that have enjoyed success in Antarctica will work in the Arctic, or vice versa, though the experience in development studies across the globe suggests that policy advocates will try to force policy solutions from one context on to other regional contexts regardless of their inherent differences. What would instead be far more useful would be to strengthen the very different existing policy instruments in the Arctic and Antarctic. This volume is shedding new light on the contrasting regional contexts of science policy in the polar regions.

The chapters in this collection represent a timely contribution to the current International Polar Year, coming as it does 50 years after the previous polar year, the International Geophysical Year (1957–58). There exists at present no satisfactory historical, political or cultural analysis of International Polar Year events – or indeed of other similar types of international research efforts like heliophysical, biological or oceanographic years. This is by no means to overlook the importance of research by historians and geographers on the history of specific polar year events. However the role of international scientific institutions to create large-scale international, synoptic research programmes every 25 or 50 years should itself be seen as part of the repertoire of science policy instruments that deserves to be the subject of a research programme in its own right.

One of the ways in which this IPY distinguishes itself from previous ones is that the inclusion of a 'humanities' theme 'to investigate the cultural, historical, and social processes that shape the sustainability of circumpolar human societies, and to identify their unique contributions to global cultural diversity and citizenship' (ICSU 2004, 7). Implicit in this theme is the notion that circumpolar societies require access to the intellectual capital that we call 'research', whether done by themselves or by others, to ensure that they can exercise their own duties as citizens to the full, and plan their own sustainable futures. Seen in that light, the recognition of northern citizens as well as social scientists as legitimate participants in this IPY represents a policy landmark and a departure from previous polar years, even if its implementation placed severe constraints on this participation. To go one step further, I would suggest that reflecting on the political, cultural and ethical conditions that have made possible international polar years in their different manifestations, represents the fullest expression of the critical enlightenment values to which they are ultimately indebted. Examining the political and philosophical structure of polar years, far from being a mere armchair excursion, is arguably a point of departure by which the humanities can meaningfully seek to articulate the legacy of this International Polar Year in relation to the tradition that it belongs, and yet transforms.

Taken as a whole this volume offers a fresh and often original set of views of the changing political entanglement between science and policy in the polar regions. The chapters are highly interdisciplinary, bringing political science and history of science together with geography and public policy analysis. The positioning of polar year research and governance in relation to arguments about the emergence of non-state actors like the indigenous Permanent Participants of the Arctic Council alongside the traditional Westphalian states (Shadian) provides the outline of a framework in political theory for developing a longue durée history of polar research. So too by attending to the importance of Cold War politics and environmentalism (Nilsson), it becomes possible to see much more clearly why the polar regions have in different ways been sites of exceptionalism, often insulated politically as well as epistemologically from wider global debates about participatory democracy. Some of the chapters are prescriptive and recommend reform to existing scientific bodies like the Scientific Committee on Antarctic Research (Jabour and Haward), while others seek to provoke by arguing the rhetorical functions of polar year events causes analysts to overlook the role of science in state-funded and private sector resource exploration (Huebert) as well as overlooking new forms of private-sector intellectual property rights (Chaturvedi and Rothwell), and more broadly speaking, bio-politics (Tennberg).

I am delighted that this book is a contribution to the International Polar Year project to study the role of polar research stations in the governance, culture and heritage of the polar regions.[2] When I first proposed a collaborative project about polar research stations (see chapters by Wraakberg and Shadian in this volume), there were very few scholars seeking to discover the fascinating and often hidden local histories of research stations. My own experience as a researcher based at the Igloolik research centre in the late 1980s had shown me how deeply the existence and functions of research stations are shaped by their political, cultural and historical contexts. However the first full-length geographical ethnography of a polar research station was undertaken by my then doctoral student, Richard Powell, in a full-length study of the historical geography of the Polar Continental Shelf Project at Resolute Bay, Nunavut. When Jessica Shadian came as a visiting doctoral student to work with my research group in 2005, it became clear in the course of many conversations that scientific institutions like research stations can be a lens through which to view the institutional structures of scientific authority, from the local level right up through national research councils to the international councils and organizations of science.

Reference

ICSU (2004), 'A Framework for the International Polar Year 2007–2008' <http:// classic.ipy.org/development/framework/framework.pdf>, accessed 9 January 2009.

---

2   Polar Field Stations and International Polar Year History: Culture, Heritage, Governance (1882–Present), IPY Project ID 100, PI: Michael Bravo, Sverker Sörlin.

List of Abbreviations

AAC	Arctic Athabaskan Council
ACAM	Australian Collection of Antarctic Micro-Organisms
ACAP	Agreement for the Conservation of Albatrosses and Petrels
ACIA	Arctic Climate Impact Assessment
AEPS	Arctic Environmental Protection Strategy
AIA	Aleut International Association
AMAP	Arctic Monitoring and Assessment Programme
ATCM	Antarctic Treaty Consultative Meeting
ATCP	Antarctic Treaty Consultative Party
ATS	Antarctic Treaty System
CAFF	Conservation of Arctic Flora and Fauna
CAML	Census of Antarctic Marine Life
CARA	Circum-Arctic Resource Appraisal
CBD	Convention on Biological Diversity
CCAMLR	Commission for the Convention on the Conservation of Antarctic Marine Living Resources
CCAS	Convention for the Conservation of Antarctic Seals
CEE	Comprehensive Environmental Evaluations
CEP	Committee for Environmental Protection
CFCs	Chlorofluorocarbons
CIDA	Canadian International Development Agency
CLCS	Commission on the Limits of the Continental Shelf
COMNAP	Council of Managers of National Antarctic Programs
CRAMRA	Convention on the Regulation of Antarctic Mineral Resource Activities
CSAGI	Comité Spéciale de l'Année Géophysique Internationale
DEW	Distant Early Warning
EEZ	Exclusive economic zone
EIA	Environmental impact assessment
EU	European Union
FAO	Food and Agriculture Organization
GARP	Global Atmospheric Research Program
GCI	Gwich'in Council International
GEUS	Geological Survey of Denmark and Greenland
IASC	International Arctic Science Committee
ICC	Inuit Circumpolar Council
ICRW	International Convention for the Regulation of Whaling

ICSU	International Council of Scientific Unions/International Council for Science
IGY	International Geophysical Year
IK	Indigenous Knowledge
ILO	International Labour Organization
IMO	International Meteorological Organization
IPCC	Intergovernmental Panel on Climate Change
IPY	International Polar Year
ISA	International Seabed Authority
IUCN	International Union for the Conservation of Nature and Natural Resources
IUU	Illegal, unreported and unregulated
IWC	International Whaling Commission
JARPA	Japanese Whale Research Program under Special Permit in the Antarctic
NATO	North Atlantic Treaty Organization
NGO	Non-governmental organization
NOAA	National and Oceanic Atmospheric Association
POPs	Persistent organic pollutants
RAIPON	Russian Association of the Indigenous Peoples of the North
SC	Sámi Council
SCAR	Special Committee on Antarctic Research/Scientific Committee on Antarctic Research
S&TS	Science and technology studies
TRIP	Trade-related Aspects of Intellectual Property Rights
UN	United Nations
UNEP	United Nations Environment Programme
UNESCO	United Nations Educational, Scientific and Cultural Organization
UNCLOS	United Nations Convention on the Law of the Sea
UNFCCC	UN Framework Convention on Climate Change
USAID	United States Aid for International Development
USGS	United States Geological Survey
VMS	Vessel Monitoring System
WIPO	World Intellectual Property Organization
WMO	World Meteorological Organization
WTO	World Trade Organization

Introduction

Jessica Shadian and Monica Tennberg

Welcome to IPY: It is envisioned that the International Polar Year (IPY) 2007–2008 will be an intense, internationally coordinated campaign of research that will initiate a new era in polar science. IPY 2007–2008 will include research in both polar regions and recognise the strong links these regions have with the rest of the globe. It will involve a wide range of research disciplines, including the social sciences, but the emphasis will be interdisciplinary in its approach and truly international in participation (IPY March 17. 2005).

On 1 March 2007 at the Palais de la Découverte in Paris, the fourth International Polar (IPY) was launched. The opening of the IPY – the largest global scientific collaborative effort to date – in many ways was also a commemoration of a long standing tradition of international polar science collaboration. Officially coordinated by International Council of Science (ICSU), the World Meteorological Organization (WMO), the Antarctic Treaty System, as well as the Arctic Council and numerous other intergovernmental and non-governmental organizations, the 2007–08 IPY brought together over 63 nations, almost 400 projects and fifty thousand scientists. The aim, building off its three predecessors, was to better understand the geophysical, biological, and for the first time – human aspects of earth's poles in advancing the frontiers of science. This fourth IPY as part of a larger polar research tradition offered a fruitful starting point to study the complexities involved in the formation of national, regional, international as well as non-state scientific interests and the construction and governance of scientific knowledge (particularly regarding the relations between humans and the environment) over time.

At its historical origins, the first International Polar Year (1882–83) was established with the intention of expanding our global understanding of some of the most intricate scientific puzzles of its time. Likewise (and central to the scientific motivations at hand), the first IPY unfolded during an international political era where expeditions to explore and exploit polar resources and intense national competition of reaching the poles first were of global prominence. Since the first IPY, the global political context has been vastly transformed including the role and perceived relevance of the polar regions in global politics. This includes a shift in the relevance of the poles as the last frontier of national conquest to the new global frontier of scientific observation. Furthermore, in the case of the Arctic, a homeland to the indigenous inhabitants who have sustained this region for thousands of years, traditional 'modern' science is being contested by the growing political agency of indigenous practice and methods for production of knowledge.

The second IPY took place in the interwar years from 1932–33 followed by the International Geophysical Year (IGY), which took place in 1957–58 at the height of the Cold War. Throughout this history, the international politics dominant during the IPY years helped structure the agendas and means of the scientific cooperation. As such, the goals and mission put forth for Arctic and Antarctic scientific cooperation during the IPYs can be construed effectively as signs of the times. The IGY, for instance, is not only remembered by the launching of Sputnik, but moreover, at a time when there was almost no political cooperation between the East and West, the Antarctic Treaty was put into place. The Antarctic Treaty demilitarized the Antarctic and created the first 'international research laboratory' to be managed by all respective states party to the treaty.

The IGY was followed by an end of the Cold War and as such the end of a political era. The perception of the Cold War as the single largest threat was soon overtaken by a new global threat through the birth of a new environmental consciousness. The implications of this new global environmental awareness coincided with the emergence of greater multilateral cooperation including the growing prominence of non-governmental actors, indigenous peoples' groups and environmental organizations with power to help define and steer the future course of global politics. Widespread concern over climate change and its impacts in polar regions has further increased interest in the poles and awareness of their problems as part of the global environmental problematique. The polar bear has recently become a popular symbol for global efforts to tackle climate change.

The histories of the IPYs have added significance to the polar regions regarding scientific research, meeting the challenges of environmental problems and achieving new forms of global environmental cooperation. The role of weather and the atmosphere, most generally, has always been of great interest to scientists, whether as a field of study in its own right or in aiding the process of territorial expansion and resource exploration. Most recently, the role of meteorology has become central to understanding the implications of environmental degradation and the processes of climate change. These new foci for meteorologists have once again turned scientific attention to the poles. Despite this renewed interest, however, the political mindset of the poles as largely uninhabited last frontiers has not submerged. Polar climate change science is being carried out in tandem with resurgent efforts to extract and exploit the excesses of the polar region's resources.

It was in this global context where the juxtaposition between the intensifying pressure towards the use of polar natural resources and scientific inquiry into better understanding global climate change (often phrased as sustainable resource management) that the fourth IPY 2007–08 commenced. In addition, the fourth IPY, for the first time formally included the human dimension in its research programme. As such, beyond the official incorporation of new non-state actors and organizations – including the indigenous peoples who live in and off the Arctic's land and resources and private companies interested in taking advantage of new opportunities which climatic changes are expected to bring out – social studies of the impacts of these changing phenomena are part of the very research plans of the IPY as well.

Uneven contours: From disciplinary bound theory to multidisciplinary issues

The chapters in this book offer a diverse set of political, legal, geopolitical and historical perspectives to the broad reach of the IPY. In the social sciences there is a growing trend towards multidisciplinary research projects that focus on particular issues and problems rather than addressing questions within the limitations of traditional disciplinary divides. This *multidisciplinary perspective* is the assumption upon which this book proceeds. Broadly speaking, this book sets out to provide a distinct contribution for those interested in the relationships between science, international politics, law and history. Most specifically, this book provides case studies and theoretical reflection of various ways in which polar knowledge is constructed, legal institutions come into practice as well as the political meaning and significance of the poles as it is played out through the IPY. The authors, coming from a multitude of backgrounds including law, history, international relations, policy, biology and journalism all focus on various aspects of the IPYs both past and present. Combining both a multidisciplinary perspective and a combination of theory, policy analysis and historical narrative this book is the first comprehensive account of its kind which focuses explicitly on the political, legal, and historical aspects of polar science through an account of the IPYs.

Polar opposites? A comparative perspective of the poles

The history and politics of the two polar regions varies greatly. The Antarctic is a frozen continent and devoid of human settlement (not taking into account the scientists who live and work there). The histories of the IGY (1957–58) and the establishment of Antarctic Treaty system (ATS) in 1959 are closely linked. The Antarctic Treaty includes nowadays several protocols to cover various environmental issues in the region. The treaty has demilitarized the region as well as made it into a scientific laboratory and area of international research cooperation.

The Arctic on the other hand is comprised of the Arctic Ocean and its limits and the very politics of its boundaries is perhaps a prelude to its complexity. The Arctic has human populations including eight states, four million people and 500,000 indigenous peoples who have varying levels of political autonomy. Since the late 1980s the region has emerged as an area for environmental and scientific cooperation, especially through the establishment of the Arctic Council in 1996. The Arctic Council is comprised of the Arctic states accompanied by six indigenous permanent participants. Since its founding further research discoveries and receding ice have renewed an ongoing debate over how best to divide the north among those with varying degree of rights and sovereignty in the Arctic.

The contents of the book

This book is broken down into two main parts. In the first part, *Whose Arctic? Constructing Arctic politics through claims of knowledge*, the role of knowledge producers with the evolution of Arctic environmental politics, especially in climate politics is described and discussed. The development of climate science and history of IPYs is studied from the point of view of polar science, the politics of sustainable development and climate change, Arctic policy and the history of Arctic research stations.

In Chapter 1, Annika Nilsson concentrates on climate change science and the interplay between Arctic and global perspectives. In the past 150 years, the scientific image of the Arctic climate has shifted dramatically. Nilsson places the development of climate change science and its interest in the Arctic into the context of these political developments globally and in the Arctic. It highlights how indigenous knowledge is becoming increasingly recognized and connects this recognition to norms for international cooperation in the Arctic and to a challenge of the exclusive prerogative of academic science to speak for nature and to define the Arctic.

In Chapter 2, Jessica Shadian looks at the IPYs as a way in which to understand global political change. Non-state institutions are increasingly part of the foundation upon which dialogue, governance and the construction of knowledge proceeds. Building on this assumption and putting the IPY into a historical perspective of Westphalian politics, Shadian focuses on the role of the science community in shifting the boundaries of law and politics; the role of indigenous political actors in transforming the basis of IPY science and the role of private industry as another instance of non-state involvement in IPY science.

Chapter 3 by Rob Huebert examines the assumption by most scientists that science is value free. This includes the reaction of the international community to the scientific discovery of different types of environmental degradation in the polar regions. Unlike past Arctic political efforts and treaties, climate change and its implied assumption of greater maritime accessibility has not brought forward an international effort to create a diplomatic solution to this environmental problem. As such, Huebert examines why there is such a different set of reactions to what are ultimately the scientific study of the environmental degradation of the polar regions. Why did the international community respond cooperatively in one instance while presently it is adopting a much more unilateral approach?

From a more historical perspective, in Chapter 4 Urban Wråkberg analyzes the politics interwoven with the creation and maintenance of polar field stations for carrying out scientific observation. Wråkberg concentrates on the diverse and sometimes contradictory motives for constructing polar stations which includes geographical sensationalism as well as the politics of territorial expansion and occupation and colonialism. With a broad political and social science perspective Wråkberg pays particular attention to the practices, scientific rhetoric and public goals of IPY enterprises as followed into the present IPY.

The second half of the book turns to the Antarctic. Part II, *Whose Environment? Science and Politics in Antarctica*, describes and discusses the developments of how territorial disagreements were settled and how international scientific cooperation was institutionalized in the Antarctica since the 1950s through the role of Scientific Committee of Antarctic Research (SCAR) as a scientific advisor in the ATS and principles of ATS as it developed into a more comprehensive regime over the years. Some of the new challenges in the ATS and science are discussed in this part as well, especially the issue of bio-prospecting and accompanying difficulties as regards the nature, ownership and use of scientific knowledge about the environment.

In Chapter 5, Marcus Haward and Julia Jabour provide insight into IPY 2007–08 through a history of the Antarctic Treaty. Haward and Jabour examine the ways in which, based on previous initiatives, the fourth IPY provided a significant new momentum for international collaboration and coordination in Antarctic science including the ways in which the role of Scientific Committee on Antarctic Research (SCAR), as a non-governmental body for the international scientific activity for the Antarctic Treaty System, has developed over the years and what new challenges SCAR has.

In Chapter 6, Donald Rothwell examines intellectual freedom of scientific research through a comparative case study of Antarctic research from a legal perspective with appropriate consideration given to diplomatic and political consequences. Rothwell addresses these challenges in the Antarctica Treaty by exploring several particular sets of questions including whether the Antarctic can or should be used and exploited in the name of science (as other parts of the world have been); whether or not different principles apply; and if recent developments create new challenges for the ATS.

Consuelo León Woppke in Chapter 7 highlights the influence of domestic political and scientific factors on Chile's contribution to the formation of the Antarctic Treaty. Woppke provides an account of the political maneuvering involved in Cold War Antarctic science between the Cold War 'superpowers' and Chile and the significance of this history for Chile's involvement in IPY 2007–08. To this day Chileans find it difficult to appreciate the corollary – or possible corollary – between scientific activities and the defense of their Antarctic rights.

In Chapter 8, Sanjay Chaturvedi turns to the politics of biological prospecting – or bio-prospecting – in the Southern Polar Region. In particular Chaturvedi focuses on the issues of achieving access to Antarctic resources within a regulatory framework capable of preserving the interest of humankind in the conservation and sustainable development. Chaturvedi also pays attention to questions regarding the technological advances and financial returns emanating from bio-prospecting and how they should be equally shared under the authority of relevant international institutions or multilateral regimes. The point for Chaturvedi is to draw out the complexities putting into place (through consensus) a sound legal–political arrangement to restrict and regulate the commoditization and commercialization of polar biodiversity.

The final chapter by Monica Tennberg concludes with a discussion of the case study chapters in order to highlight the relationship between power and knowledge in polar sciences. Using Foucault, Tennberg points to the legacies of scientific and political cooperation as effects of power relations and their workings: the polar regions are established as 'scientific laboratories' and part of global 'environmental panopticons', a system of surveillance for environmental research highlighted in the tradition of IPYs in polar regions. However, as Foucault suggests, power relations are always challenged, and effects of power are contested in various ways. In the polar regions, new ways of understanding and using knowledge questions the existing modes of governing and their knowledge base.

The culmination of these ten narratives brings to bare the broad and complex political, historical and legal aspects of the IPY and the more general relationship between science and politics. In addition, these chapters draw out the way in which the practices and meanings of science changes with political and social change, including the increased authority of non-state actors such as indigenous communities as scientific experts for creating and controlling the flow of intellectual knowledge, as well as the ongoing shifting meanings of polar field stations and research institutes for defining and producing legal norms and policies. In sum, this book sets out to fill an ever growing niche in efforts to bridge various disciplines to study certain phenomena by providing a unique account of the IPY; an instance which reflects the intersection between states, non-state actors and international regimes, and the ways in which knowledge is constructed and used in changing perceptions and politics of the polar regions. The overarching research question guiding this book sets out to explore is: What is knowledge? Who knows the polar regions? Who governs polar scientific knowledge? Finally, how has the power and authority over this knowledge shifted over time and how are these changes reflected through the IPY?

Reference

Beland, M. and Allison, I. (2005), 'Welcome to IPY', *IPY* [Online] <http://classic.ipy.org/about/>, accessed 18 July 2008.

PART I
Whose Arctic?
Constructing Arctic Politics through Claims of Knowledge

Chapter 1

A Changing Arctic Climate:
More than Just Weather

Annika E. Nilsson

Introduction

The history of the international polar years have been described as a mirror of progress in science and society with an emphasis on the increasing coordination of data gathering and analysis (Behr et al. 2007). It is also a mirror of major changes in political ideologies affecting scientific collaboration over the past century and a half. This chapter is a story of how political developments have influenced the international polar years as seen through the example of climate change science and its interest in the Arctic.

Since the first Polar Year in 1882–83 the change in the structure of international society has been immense. At the end of the nineteenth century, nationalism was still the dominant ideological ideal and colonialism a central structuring principle in international society. While state-sponsored colonialism had dwindled during the latter part of the twentieth century, an onus on the nation state as the only legitimate actor in an anarchic international system was still alive well until the 1970s (Morgenthau and Thompson 1993; Bull 1977; Waltz 1979). This included a strong emphasis on self-interest and military power as guiding principles and a division of the world into East and West after the Second World War (Dunne and Schmidt 2001).

Beginning in the 1970s, a surge in oil prices, calls for a new economic world order from previously colonized countries, and a growing awareness of transnational pollution made it clear that nation states were more interdependent than previously recognized (Keohane and Nye 1994). From a need to handle the increasingly transnational politics in an anarchic structure grew an interest in international collaborative efforts. However, conflicts between the global North and the global South were also coming to the fore, not least in relation to the rise of environmentalism on the international agenda in connection with the United Nations (UN) Conference on the Environment in Stockholm in 1972 (Selin and Linnér 2005; Linnér and Jacob 2005). These conflicts have remained central to international governance as it has grown increasingly complex with both vertical and horizontal interplay among environmental and other regimes (Young 2002). Moreover, states are no longer the only legitimate actors, giving space to an increasing role for nongovernmental organizations and other transnational actors, not least indigenous peoples as has been apparent in the Arctic. At the same time, the distinction between national and international society has become increasingly blurred and the complexity of the political architecture is captured in analytical concepts such as Earth system governance (Biermann 2007) and 'glocal' (Gupta et al. 2007). Scientifically, the emphasis is on nature and society as intimately linked to a point where human society is seen as a major driving force for global environmental change (Steffen et al. 2002).

This chapter argues that scientific interests in the Arctic and the international polar years mirror these changes in international society and looks at how interplay between scientific, practical, and political motives has moved Arctic climate science in new directions over time. Based on the analytical concept of co-production of science and policy (Jasanoff and Wynne 1998; Jasanoff 2004b) and ideas from actor network theory (Latour 1987), it highlights the close connection between the developments of climate science and the international political landscape and how it has played out in the framing of Arctic climate change. The chapter covers the time period from around the first IPY to the presentation of the Arctic Climate Impact Assessment (ACIA) in 2004 (For an in-depth discussion of the interplay between science and political development as it plays out more recently including the latest IPY (2007–08), see Shadian; Rothwell in this volume).

From nationalism to emerging international networks

The connection between global warming and an interest in the Arctic goes back to a major scientific controversy in the mid-1800s: the riddle of the ice ages (Weart 2003, 5). The key question then was what could have caused the ice sheet that left scraped bedrock and landscape features that resembled those seen close to glaciers in the Alps. Could changes in the atmosphere possibly lead to such cooling that the ice could begin to grow and later melt again? A French physicist – Joseph Fourier – had already shown that the atmosphere could trap heat much like a 'hot house' or what later became labeled the 'greenhouse effect'. The scientific community involved in the nineteenth century ice-age debate was also aware that snow and ice covering a region during an ice age would reflect sunlight back into space and keep it cool, which in turn could change wind patterns and ocean currents in ways that would cool the region even further (Weart 2003, 5). This curiosity about the role of ice was the scientific context in which Svante Arrhenius wrote his famous 1896 paper on the influence of carbon dioxide on Earth's temperature, which links the burning of fossil fuels to global warming (Arrhenius 1896). The intellectual environment of the Stockholm Physical Society in which Svante Arrhenius was active also created early links between climate science and interests in the Arctic (Crawford 1998, 23). An account of how 'the discovery of the ice age' had a critical role in 'research desire' concerning the Arctic illustrates how the glacial theory – that the northern hemisphere had been covered by ice – had been a major inspiration for the first Swedish polar expeditions in 1858 and 1861 (Karl

Chydenius' account of the Swedish polar expedition in 1861 as cited in Frängsmyr 1982). Arrhenius' feat was to also include in his analysis the emerging knowledge about the carbon cycle that had been developed by his Stockholm colleague Arvid Gustaf Högbom.

The early connection between the Arctic and climate change science were made at a time when nationalism was still the major political context of polar research. Interests in the Arctic were part of the colonial history of the region and the use of northern imagery bolstered a national identity. Stockholm at the time of the early polar expeditions was what Latour (1987) calls a 'center of calculation', in this case of a national colonial science where images of the Arctic were brought from the colonial peripheries to be presented in the metropolis (Bravo and Sörlin 2002). In contrast to the present, this early climate science was not part of any internationally coordinated effort to understand the Earth as a system. The impetus for international coordination came from more immediate concerns, namely a need to predict weather.

International collaboration came early to meteorology with the first meteorological network dating back to the mid-1600s and international networks based on instrumental readings dating back to 1780 (Serafin 1996; WMO 2005c). However, the lack of simultaneous observations at several places hampered efforts to predict the weather and the real breakthrough came only with the development of the telegraph, which could be used to share data and to distribute forecasts. Realizing this new potential, several European countries established central meteorological offices in the latter half of the nineteenth century (Nationalencyclopedin 1994; Holmberg 2004). The collaboration among these offices illustrates a beginning shift in the context of climate science towards building professional networks across national borders that could connect the national centers of calculations making them less isolated within national ideologies. Or using Latour's terminology, the centers of calculation are starting to connect to each other in networks (Latour 1987). In Sweden, one of Arrhenius' colleagues at the Swedish Physical Society used the development in Europe as an argument to reform the Swedish meteorological central office to make weather forecasts available for fisheries and shipping (Holmberg 2004). A prerequisite for the new weather-forecasting activity was to ensure some kind of international coordination of measurements, which led to the creation of the International Meteorological Organization (IMO) in 1873. The IMO was a non-governmental organization but with time, this initially fragile actor network became increasingly formalized. One of its early activities was to organize the first IPY 1882–83 (WMO 2005a).

The notion of a polar year was based on a shift in the ideal about how polar research should be conducted, a shift from nationalism to internationalism. Karl Weyprecht, who originally proposed the idea of a polar year, thought it was time to collaborate internationally rather than continue the independent expeditions aimed at geographical exploration but with limited scientific value (Barr 1982). With the IMO in place, there was an international platform from which to put his ideas forward. The different expeditions and auxiliary stations that were launched during the first IPY gathered vast amounts of data. However, the data were initially published in different languages without any coordination and it appears that the political context was not quite ready to fully support a shift of polar research to an international endeavor. Greenaway (1996) has described it as an affair, on the whole, of individuals backed by national academies. The lack of coordination in this early internationalist effort can also be understood in the context of the general development of science at that time. As Crawford describes, there was a surge in international scientific congresses and organizations at the end of the 1800s, made possible by a new mode of communication: the railroad. In some instances legacies were created, especially in creating standards, while other endeavors were more short-lived. Crawford points out that nationalism and internationalism in science coexisted to a degree that had not occurred beforehand and was not to recur again (Crawford 1992, ch. 2).

The remaining nationalistic political context of science was also apparent in the Arctic, not least in relation to the vast resources of the region. Even some early examples of international cooperation in the Arctic, such as negotiations to regulate nature conservation and exploitation of natural resources on Spitsbergen, were guided by economic and ideological nationalistic interests (Wråkberg 2006). However, there are some early signs of the Arctic being framed in relation to its global role, such as the efforts to protect musk ox in Greenland and reindeer on Spitsbergen, that were part of a wider movement of world-wide nature protection (Young and Osherenko 1993; Selin and Linnér 2005).

The first half of the twentieth century brought two world wars that put a damper on international collaboration both in the Arctic and in the world at large. However, the idea of cooperation in polar research managed to remain through the interwar period. As early as 1925, scientists started discussing a second international polar year. The economic situation forced the enthusiasts to postpone their ideas and when it was eventually launched, it was much more modest in scope (Liljequist 1982). In the mean time, breakthroughs in aviation brought an increasing need for better weather forecasts. Thus, when the second IPY was eventually realized in 1932–33, the IMO had a special focus on improving weather observations in the polar region. Greenaway (1996) has described the second IPY as initiated and energized by a body concerned with a limited subject. Mirroring some of the changes that had taken place in scientific international cooperation in the previous half-century, the second IPY initiative was more international than the first IPY.

In summary, the early development of climate science and its relation to polar research shows how individual scientists and scientific networks with an interest in ice ages became increasingly organized and international in their orientation. The development was helped along by technological breakthroughs in communication, such as the railroad and telegraph, and also by society's needs for better weather forecasting. In the Arctic, emerging international scientific networks were driven by professional organizations rather than by states. This mirrors an international society where nationalism was still the fundamental structure and where international regimes had yet to become important for coordinating state activities.

While cooperation among meteorological offices resulted in the creation of the IMO, the polar efforts did not leave a pan-Arctic organizational legacy, possibly because interests in the Arctic were still mainly related to nationalistic interests. The region was peripheral to central colonial powers within the nation states with Arctic territories. Arctic developments are therefore best understood in the same light as the global development at the time, where the emerging international cooperation was an instrument for colonial powers to assert their interests in their territories, whether it concerned threatened species or economically important resources (Linnér 2003, 30). The Arctic situation is also compatible with observations that internationalist developments in science from the late 1800s to the 1930s did not leave the nationalistic tendencies of science behind. Even if better communications allowed new international collaboration, this was also a time when the ideology of nation-building was strong and where science was part of this effort (Crawford 1992, ch. 2).

International cooperation and global environmentalism

The end of Second World War signaled a new era in international society with a growing focus on international cooperation. In the Arctic, Cold War military interest delayed this development until the early 1990s. This era also feature growing interactions between science and policy, where scientists entered as new actors on the scene of international relations. Illustrated by the IGY 1957–58 and the rise of international collaboration in climate science, the 1950s and 1960s featured an emerging global gaze in science. The rise of environmentalism in the 1960s and 1970s, along with the challenges of transnational pollution, brought renewed interest in international political collaboration. In both the scientific and political realms, reaching out across national borders became a necessity that with time developed into a norm.

The idea of an international geophysical year was born by a small group of US scientists in 1950 from an informal discussion on how new technology from the Cold War could be used for advancing the science of geophysics (Greenaway 1996; IPY 2007–08 2005). Their ideas fit well into US Cold War military ambitions. This would be a way of collecting data in areas that would otherwise be inaccessible. In addition, the effort could be used to gain prestige in the ideological war between East and West. Some also saw the potential for using scientific cooperation as a means to thaw East–West relations (Weart 2004). The non-governmental scientific organization International Council of Scientific Unions (ICSU) was the linking point for IGY. Compared to the previous international polar years, the initiative had thus moved from the meteorological community with its focus on weather observations to a community interested in more fundamental aspects of Earth as a planet and to an international body concerned with science at large (Greenaway 1996). The responsibility for the meteorological research program within IGY was placed with the World Meteorological Organization (WMO). The WMO had grown out of the IMO and become a special agency under the UN in 1951 (WMO 2005b).

The creation of the WMO exemplifies the political context of and the drivers for the IGY and climate science during the post-war years. According to Miller (2001), the development of the WMO was intimately connected to a US foreign policy of scientific internationalism and ideals to construct a stable world order that could serve American interests. The WMO was an ideal model for post–World War II US foreign policy because it could build on existing network among the national meteorological offices at the same time as the new regime could be framed as a matter of technical cooperation that would not threaten the sovereign rights of other states (Miller 2001, 177). The WMO mustered new economic resources, which were used to invest in new technologies that improved the capacity to observe the Earth's weather. It also committed governments to standardize the measurements (Miller 2001). For the IGY, the WMO was important because, in the words of Miller, the 'prior existence of international data collection networks and protocols, authoritative bodies capable of setting priorities, and institutional support for the publication of data substantially eased the burden of building a globally coordinated research enterprise' (Miller 2001, 200). The WMO was also instrumental in shifting IGY's focus from polar to geophysical and from regional to global. The two polar regions were now viewed as integral to the world's climate system (Greenaway 1996).

For global climate research, the IGY was a breakthrough because it provided an impetus to create physical infrastructure that later became critical to observing climate change. One example is the chains of weather observing stations that were established to collect data simultaneously and the fact that all the meteorological data were collected in three central World Data Centres (Greenaway 1996; Edwards 2001). The IGY also marked the initiation of regular measurements of carbon dioxide levels in the atmosphere. The data from Mauna Loa, Hawaii, displaying a curve of steadily increasing concentrations have become an icon to show the influence of human civilization on Earth's atmosphere. Weart describes this as the 'capstone' on the structure built by early climate scientists and the 'discovery of the possibility of global warming' (Weart 2003, 38). Logistically, the IGY pushed Earth scientists to coordinate their work to a greater extent than before, creating the foundation of an international scientific community (Weart 2004).

During the IGY, the Arctic was not in focus the same way as in the previous polar years. Instead, emphasis was placed on the Antarctic, which until this time had been inaccessible to any large-scale scientific investigations. In the Antarctic, the IGY brought the potential aspects of international science collaboration to the fore initiating what became the Antarctic Treaty (see Jabour; Haward in this volume).

The IGY also became a starting point for looking at planet Earth as a whole. In particular, the Soviet Sputnik satellite in 1957 together with later American space launches marked the beginning of the space race (Weart 2004, International-4). Several writers (Bryld and Lykke 2000, 10; Jasanoff 2004a, 37) have pointed out that the changed awareness brought on by looking at Earth from the outside has been fundamental for our culture, similar to the scientific revolution. Previous to this, the world as a whole could only 'be seen in the imagination' (Greenaway 1996, 155). The scientific imagination and networking that led up to the IGY can thus be seen a harbinger of a new cultural era (see also Shadian in this volume). With these scientific, political, and cultural developments to set the stage, a new issue was emerging on the political agenda: the environment.

The impetus to start thinking about environmental impacts also came from a war technology: the atom bomb. Atmospheric tests of atom bombs had made it apparent that the impacts could be geographically far-ranging. It was suddenly possible to think that human society could be a threat to itself (Linnér 2003). In the United States, protests against nuclear tests became an early embryo of the environmental movement, to which Rachel Carson could appeal when she was protesting against the use of chemical pesticides in her seminal book *Silent Spring* (Worster 1994; Lear 1997). In the Arctic, the nuclear tests became a rallying point for local control of the environment when indigenous peoples in northern Alaska managed to stop the use of nuclear explosives to excavate a harbor. Moreover, Alaskan indigenous people were concerned about how the fallout from nuclear tests would affect reindeer and caribou, sparking an initial awareness about contaminants in indigenous traditional foods (Hild 2004).

Climate change was not yet a major public or political issue in the late 1950s and early 1960s, but it was recognized as a potential problem. A popular belief was that fallout from the bombs could cause climate cooling. Such fears, voiced by both scientists and politicians, prompted US federal nuclear programs to investigate dispersion of fallout in the atmosphere and later also to institutionalize research on climate change (Hart and Victor 1993). Based on calculation published by US carbon cycle scientists Roger Revelle and Hans Seuss already in 1957, climate change had become framed as 'a large-scale geophysical experiment' (Revelle and Seuss 1957). Revelle regularly advised US federal agencies and in a 1965 report to the US president, where he chaired a subcommittee on the atmosphere, it was officially recognized for the first time that climate warming could be caused by human activities and have important consequences (Kellogg 1987; Hart and Victor 1993; President's Science Advisory Commission 1965 as quoted in Agrawala 1998a; Weart 2003).

Similar to the setting for climate science in the early twentieth century, Revelle and his scientific colleagues were active in a tension field between nationalistic interests and internationalism. Although Revelle played a key role in the planning of IGY and is an example of the emerging role of climate scientists as political actors, he was also part of a US scientific élite that managed to gather resources for carbon cycle research because it could take advantage of US Cold War military interests (Hart and Victor 1993; Doel 2003). Doel (2003) has highlighted the close links between US geophysicists and US national security interests and the need for military secrecy to explain why carbon cycle scientists at the time failed to connect their results to emerging research on climate modelling. They also failed to make the growing knowledge on carbon cycling an international issue (Hart and Victor 1993).

The breakthrough for *international* climate science instead came via a wish to use the new satellite technology to gather data about the weather conditions high in the atmosphere, creating a basis for better weather forecasts, including the first generation of computer models for weather prediction. There are thus some similarities to the nineteenth century drive for better weather forecasts but with new technological possibilities and in a political context where internationalism had gained political momentum. In 1963, in an address to the United Nations about the peaceful uses of satellites, US president J. F. Kennedy proposed an international cooperative effort for weather prediction and eventually weather control in order to have better understanding of the global climate system (Greenaway 1996; Agrawala 1998a; Edwards 2001; Bolin 2007). By this time, the WMO was up and running, which provided a suitable international organizational framework for such a proposal. In 1963, the World Weather Watch became one of its core activities (Weart 2004; Bolin 2007). A major activity for the World Weather Watch was to set standards for how weather data was to be gathered. This had been a long-standing ambition among meteorologists but had been previously met with limited success (Edwards 2004). In the early 1960s, emerging changes in international society – with the UN and a growing emphasis on cooperation – had changed the conditions. Edwards describes these standards as socially constructed tools: 'They embody the outcome of negotiations that are simultaneously technical, social and political in character' (Edwards 2004). The new setting made a platform for negotiation available. This was combined with US foreign policy interests to promote internationalism in science as a way to ensure access to data from other countries and to create a new world order (Doel 2003; Edwards 2001).

In addition to the practical concerns of weather forecasting, there was an increasing scientific interest in using satellite technology to study the global climate system. As described in detail by Bolin (2007), initiatives in the WMO here joined with interests of ICSU leading to a joint effort to form the Global Atmospheric Research Program (GARP) in 1967 (see also Greenaway 1996). A central activity of GARP was to gather data sets on a global scale, which further enforced the standardization efforts of the World Weather Watch. A driving force for standardization was that data had to be usable in computer models of the global atmosphere (Weart 2004; Edwards 2001; Bolin 2007). As indicated by the name – *Global* Atmospheric Research Program – climate science was at this time increasingly framing the issues in global terms.

To summarize the post–World War decades, formalized governmental and non-governmental cooperation created networks of people and technologies that emphasized the Earth as a system. The scientific starting point was no longer at the local or national levels but now at the global level. Research with global emphasis gained momentum from US military interests and technology and a US political push for internationalism in science. Even if the IGY of 1957–58 was partly fuelled by US nationalistic interests, it can still be seen as a harbinger of a new era of international network building in the geophysical sciences. In relation to climate change research, the international aspects became especially emphasized in the creation of the World Weather Watch and the Global Atmospheric Research Programme. The drivers for internationalization were practical, scientific, and related to the geopolitical situation all at the same time. There was a wish to understand fundamental geophysical principles by using tools that were developed by the military, which linked to political aspirations and organizational resources connected to an ideology of using science collaboration as part of creating a new world order. The previously fragile networks were strengthened by new resources, technologies and norms about international standardization and centralization. The centers of calculation became international.

The organizational development of international climate science was accompanied by a change in awareness about the environment that had also been brought forth by the war technologies. Thus the global gaze of science went along not only with political internationalism but also with the dawn of a cultural era that encompassed ideas of both the destructive power of human technology and viewing the Earth as a whole. The stage was set for a discourse on global environmental change to emerge.

The Arctic in the post-war era

The Arctic did not play a prominent role in the emerging global climate science networks during the post-war decades. There was no scientific reason to ignore the northern polar region and US national research programs were initiated because of military interests (Doel 2003). However, for international research collaboration, the Cold War political setting created major obstacles. The region was divided between the North Atlantic Treaty Organization (NATO), led by the United States, and the Warsaw Pact, led by the Soviet Union, with Sweden and Finland as non-aligned nations squeezed between the two superpowers (Young and Osherenko, 24, 190 ff). Both land and sea were heavily militarized, including radar stations for early warning of long distance missiles and tests of nuclear weapons in the Aleutian Islands and on Novaya Zemlya (Heininen 2004, 207–08). Heininen has written that the Cold War had 'transformed the region first into a military flank, then a military front or even a "military theater"' (Heininen 2004, 218). Many proposals for arms control and confidence building were put forth but the formal agreements did not include the Arctic.

Arctic military build-up was soon followed by conflicts over contested marine territories through claims for larger zones of exclusive rights for economic exploitation, as seen by the so-called fishing wars or cod wars (Heininen 2004, 218). The extensions of sovereignty claims allowed even more freedom for militarization. The same year as the UN Convention on Law of the Seas (UNCLOS) was signed – 1982 – a Swedish diplomat wrote that the Arctic was one of the most heavily militarized areas in the world and that the Law of the Sea and new zones of sovereignty made areas inaccessible to research that had been open only a decade earlier. For example, the emerging new regime concerning the high seas greatly hampered a Swedish polar expedition in 1980, when the Soviet Union would not allow sampling on its northern continental shelf (Johnson Theutenberg 1982). This setting for research collaboration can be placed in contrast with Antarctica, where the IGY had led to an international institution to coordinate research: the Scientific Committee on Antarctic Research (SCAR).

The Arctic was not devoid of cooperation. For example, indigenous peoples in the region started to organize across borders with the creation of the Saami Council in 1956 and the Inuit Circumpolar Conference in 1977. The Nordic countries also had a cooperation focusing on the northern regions of the Nordic countries, while UNESCO's Man and Biosphere Programme had established a Northern Sciences Network (Archer and Scrivener 2000, 602). However, the general picture for the Arctic differed from the emerging global cooperation as exemplified above with the case of climate science. While developments in climate science gained momentum from internationalism and Cold War politics on the global scene, geopolitical and security concerns hampered cooperation in the Arctic where the two superpowers stood face to face.

Sustainable development in 'One World'

If the post-war era of the 1950s to early 1970s set the stage for international cooperation in climate research, this period was followed by changes in international society that brought yet another dimension to climate change science and to polar research: a discourse on sustainable development. This international discourse provided new room for environmental scientists, including climate change and Arctic researchers, as advisors to policy makers. With time it has also increasingly influenced the scope and nature of climate change and polar science, including the framing of Arctic climate change. The 1972 UN Conference on the Environment in Stockholm signaled the start of this new era.

The Stockholm Conference started as an initiative by Sweden to strengthen the UN, which at the time was plagued by East–West tensions and issues surrounding decolonization. The environment was seen as a suitable theme and the result was a Swedish initiative for the UN to organize a conference on the environment (Selin and Linnér 2005). The initiative gained further salience for Swedish policy makers when it became clear that sulfur emissions from the burning of fossil fuels in other parts of Europe caused acidification of lakes and fish deaths in Sweden. It became important to convince other governments that the environment and pollution were transnational issues (Lundgren 1997, 288; Selin and Linnér 2005, 19). The UN Conference on Human Environment marked the beginning of the environment as a global political issue. As described by Selin and Linnér, the Stockholm Conference was closely linked to a political context that included an emerging international discourse on environment and development, which connected ecological deterioration, a growing demand for natural resources, and a focus on human development. Moreover, a growing recognition of the interdependence of nation states had also caused an increasing political focus on cooperation across national boundaries, even if this was also challenged in the name of protecting national sovereignty (Selin and Linnér 2005).

A major outcome of the Stockholm Conference was the creation of the United Nations Environment Programme (UNEP). This new agency was given the role of monitoring and assessing the quality of the environment and alerting the world to any environmental danger signals (Andersen and Sarma 2002). One of the first issues UNEP tackled was the threat to the stratospheric ozone layer, eventually leading to the Vienna Convention on the Protection of the Ozone Layer in 1985 (Litfin 1994; Clark et al. 2001). Only a few months after it was signed, and when a scientific assessment report was in the final stages of preparation, the British Antarctic Survey reported an unexpected, sharp depletion of ozone over Halley Bay, Antarctica. The Antarctic 'ozone hole' caught the attention of scientists, policymakers, and the public alike. Not only did this discovery speed up negotiations for binding protocols to the framework convention, it also created a public awareness that human society could change the atmosphere in vivid enough terms to make front page news (Litfin 1994; Edwards 2001).

The ozone hole shows how the new political context of international environmental cooperation allowed a scientific discovery in a remote polar region to play directly into political decision making. According to Litfin, a group of scientific knowledge brokers were able to control, frame, and interpret this new information and their close interaction with the policy sphere paved the way for political regimes to be effective in reducing the use of CFCs and other ozone-depleting chemicals (Litfin 1994). In addition, one could argue that the new context of international environmental cooperation provided a setting that allowed the Antarctic ozone hole to become a bellwether for global environmental change. Not only did UNEP provide an arena for negotiations, it also initiated assessments of the scientific knowledge on which to base policy decisions thus pushing for consensus and supporting a political legitimacy that individual scientists may not have had. The situation can be placed in contrast to early warning in the United States in the 1960s about climate change where there was no international platform to carry the issue forward. The Arctic did not play a role in the early ozone discussions. A major reason was that the extremely cold temperatures in the upper atmosphere that contributed to the large ozone hole over the Antarctic were thought to be unique to the southern polar region. Arctic ozone holes are better described as Swiss cheese and are often caused by dynamic air movement (AMAP 1997, 165).

While political attention was on the ozone issue, climate science focused on improving the understanding of the global carbon cycle and how humans may affect it (Bolin 2007). Relying on the global networks for data collection that had been set up in during the IGY, the development of computer technology created new ways of getting a picture of what might be in store. But the major developments were neither scientific nor technological. They were at the interface between science and policy, building on a wish to bring new scientific findings to the attention of policy makers. In short, scientists during this time started to synthesize their understanding of the global climate systems and to phrase their conclusions as direct messages to political decisions makers, most notably in a series of meetings in Villach in the mid-1980s (Agrawala 1998a; Bolin 2007). The new recognition of climate change as a major environmental issue eventually led to the creation of the Intergovernmental Panel on Climate Change (IPCC) in 1988. At the heart of these developments was the issue of how climate change had now moved from being mainly a scientific concern to an *intergovernmental* affair (Franz 1997; Agrawala 1998a; Agrawala 1998b; Bolin 2007). IPCC's first assessment was published in 1990 and became an important input in the discussion about a global climate convention, which was realized in the UN Framework Convention on Climate Change (UNFCCC) signed at the UN Conference in Rio de Janeiro in 1992.

What role did the Arctic play in these formative years of the climate science and policy interface? Politically, the Arctic was not a recognized region and, as outlined above, international relations were colored by Cold War political dynamics. In relation to climate science, the picture of the Arctic depended on its role in the global climate system. For example, the report that was written for the seminal 1985 Villach meeting discusses the role of reflections from the polar ice in relation to climate models, the role of ice core data for determining atmospheric levels of carbon dioxide and methane and water-saturated tundra as a source of methane. Sea-ice extent is mentioned as a moderately high priority for monitoring (Bolin et al. 1986). The report's discussion of the impact of climate change focuses on agriculture and forestry and does not include consideration of the Arctic. The first assessment from the IPCC also looked at the Arctic from a global perspective but also shows an emerging interest in the sensitive ecosystems of the Arctic and infrastructure as it relates to exploitation of its resources. It mentioned that polar regions might warm two to three times more than the global mean and that some models predicted an ice-free Arctic with profound consequences for marine ecosystems (Houghton et al. 1990; Tegart et al. 1990).

During the 1990s, the IPCC continued to evolve as the central international mechanism for assessing knowledge about climate change, but there were also parallel scientific developments that have come to frame the view of the Arctic. Most important was the formal birth of Earth system studies with the International Biosphere–Geosphere Programme in 1996 under the auspices of ICSU and the smaller social science counterpart the International Human Dimensions Programme. This became a starting point for international scientific collaborations that have emphasized the Earth as a system and that this system is in a state of change (Steffen et. al. 2002).

In the 1990s, political developments in the Arctic started to catch up with the larger changes in international politics, and if Arctic scientific cooperation was stifled by military and strategic concerns during the Cold War, the late 1980s and the 1990s signaled a new era for this part of the world. But the story really starts in the southern polar region where the IGY had led to formalized international research cooperation in the SCAR under the auspices of ICSU. While, similar research cooperation in the Arctic was prevented by the Cold War tensions, the wish for pan-Arctic research cooperation remained alive within the polar research community and among non-governmental scientific networks (Interview Anders Karlqvist, 21 June 2005). Therefore, scientists with an Arctic interest were quick at picking up signals from the new political developments in the Soviet Union in the mid-1980s – President Mikhail Gorbachev's mission towards glasnost and perestroika. For example, in connection with a meeting in 1986, there were informal talks about these new signals and, according to one account, the participants agreed that it was time for the Arctic countries to start discussing such a possibility (Interview Odd Rogne, 14 April 2004). Three scientists later prepared a report that laid the foundation for further work towards a formal collaboration (Archer and Scrivener 2000). Schram Stokke has described the political circumstances in terms of softening national competitive interests in the area (especially military), which made it easier to take environmental concerns and a desire for scientific cooperation into consideration. Also, science and the environment were issues where national interests were not in such direct conflict, therefore providing maneuvering room for entrepreneurial groups such as the scientific networks around polar research (Schram Stokke 1990, 62).

Politically, many writers have pointed to a symbolic turning point away from Cold War frostiness in pan-Arctic relations when Gorbachev gave a speech in Murmansk in October 1987 (Young 1998; Archer and Scrivener 2000; Heininen 2004). The direction of his speech was to establish a program of international cooperation that included resource development, scientific research, environmental protection, opening up of sea routes, and recognition of indigenous peoples' rights (Young 1998, 32). Gorbachev also offered a new definition of the Arctic that allowed the non-rim states Finland, Iceland, and Sweden into an Arctic cooperation (Keskitalo 2002). Up until then Arctic countries were usually defined as Arctic rim states, i.e. countries bordering the Arctic Ocean (Canada, Denmark, Norway, the Soviet Union, and the United States). Many of the discussions about Arctic cooperation took place among research administrators, but it appears that they acted with the backing of, and through consultations with, their respective departments of foreign affairs (Interview Odd Rogne, 14 April 2004). The discussions, along with the new Soviet attitude, led to 'increasingly politicised exchanges first at the scientific and then at diplomatic levels' and eventually the creation of the International Arctic Science Committee (IASC) in 1990 (Archer and Scrivener 2000, 603). The main aim of IASC has been to increase knowledge about Arctic processes. Judging from documents created in its early life as an organization, there was an emphasis on the global significance of changes in Arctic climate, weather, and ocean circulation (1990 Council Meeting Report as quoted in Archer and Scrivener 2000, 604). For the first time, the image of the Arctic as a linchpin for global change appears. It appears that the focus on global processes that had been established during the IGY and carried over into Antarctic research was now moving into the Arctic via the international scientific networks that had been established with help of ICSU and the WMO. The natural starting point for scientific analysis was no longer the Arctic as part of a nationalistic image but the Arctic as an important part of a global system. The Arctic as a region defined by those living there was yet to appear.

Parallel with Arctic scientific cooperation was a surge in political diplomatic activity which culminated in the Declaration on the Protection of the Arctic Environment in 1991, the Rovaniemi Declaration, and the creation of the Arctic Environmental Protection Strategy (AEPS) as a forum for collaboration around transboundary environmental issues (Young 1998). Scientific networks may have played more of a role in the initiation of the AEPS than is apparent from Young's account of the formation of this regime, but it did not operate as an epistemic community presenting a common solution or perspective to a problem. Rather, the scientific community may have provided a network that included the Soviet Union, which could be used to scout out the potential for political negotiations.

To summarize, the two decades starting with the Stockholm Conference in 1972 was an era when concern for the environment entered international politics and when the internationally organized scientific community increasingly framed the environment as a global issue, including the birth of Earth system science. The Arctic lagged behind global development but in the late 1980s and early 1990s a new era of cooperation dawned. As had been the case outside the Arctic, scientific cooperation and environmental protection provided platforms that were viewed as sufficiently neutral to overcome remaining security or sovereignty concerns. The network of polar researchers played a key role in the initiation of Arctic scientific and political cooperation. This network was part of the legacy of the IGY but also supported by a growing institutionalization of international cooperation in climate science. The world had changed while the northern polar region remained in the grip of the Cold War. When scientists again were able to focus on the Arctic, it was with a global view.

Human dimensions enter the picture

While the initial climate science interests in the Arctic focused on the physical environment and its role for the global climate system, the growing discourse on sustainable development and increasingly close links between policy and science during the 1990s created space for new issues to enter the framing of Arctic climate change. IPCC's assessments provide an illustration. For example, in the 1997 special report *Regional Impacts of Climate Change: An Assessment of Vulnerability* (IPCC 1997), the human dimension was much more visible, with a focus on the impact of climate change on communities and indigenous peoples. The Summary for Policymakers shows the new focus: 'Although the number of people directly affected is relatively small, many native communities will face profound changes that impact on traditional lifestyles' (IPCC 1997, 8). However, the relationship between the polar regions and the global climate system remained a major concern, specifically ocean circulation and feedbacks connected to ice dynamics. When the IPCC presented its Third Assessment Report in 2001, the polar regions were granted their own chapter by Working Group II, where the analysis was developed further. In addition to topics on changes in the physical and biological environment there was also a focus on the impact on human communities with discussions about indigenous peoples and economic activities; the latter including oil and gas extraction, building, transportation, pollution, fisheries, and reindeer husbandry.

While the global climate science community was highlighting the vulnerability of the Arctic region climate was somewhat surprisingly not a key issue in the emerging political cooperation within the Arctic. Even if climate change and the effects of stratospheric ozone depletion were identified as significant threats to the Arctic environment in the early work of the AEPS, the main responsibility for measuring the causes and effects and to understand the processes was placed on other international groupings and other fora (Nilsson 2007). Climate change was thus seen as a second priority. In the first major scientific assessment of environmental threats in the Arctic, climate change and UV were included but were not at the center of attention (AMAP 1997; AMAP 1998).

In September 1996, the AEPS was subsumed into the Arctic Council. The structure of this high-level policy forum is quite radical in relation to other international bodies. Although intergovernmental in nature, it also includes permanent participant positions for the indigenous peoples of the region. The aims are wider than those of AEPS and includes promoting sustainable development (Archer and Scrivener 2000, 613). Although the Ottawa Declaration that established the Arctic Council as a high level forum for cooperation is silent on how countries were to be represented, the responsibility shifted from Departments of Environment to Departments of Foreign Affairs. Archer and Scrivener described this as foreign ministers seeing the Council as a mechanism to reassert their control over Arctic cooperation (Archer and Scrivener 2000, 615).

In summary, the institutional developments in the Arctic in the late 1980s and during 1990s set the stage for the Arctic to be a region in its own right (e.g. Young 1985; Young and Osherenko 1993; Young and Einarson 2004). It also set the stage for the Arctic to re-enter climate change science from a new position.

Climate on the Arctic agenda: The ACIA

While climate change was not a priority of the early Arctic political cooperation, it remained an issue for scientific cooperation. For example, in the mid-1990s, IASC initiated two subregional climate impact studies – the Bering and Barents sea impacts studies. Via individual scientists active in the global climate science community, IASC also had links to the IPCC and a wish in the IPCC to get better knowledge of climate change at the regional level. Key people included Bert Bolin, founding chair of the IPCC and then vice chair of IASC Executive Committee, and Robert Corell, at the time US representative and chair of IASC's Regional Board. In the global political context, the Arctic was of particular interest as a potential showcase of anthropogenic climate change – the bellwether that was needed in the climate policy debate where the scientific basis for anthropogenic climate change was still a major political issue in the late 1990s. The Arctic was also attractive for an in-depth study because the Arctic Council provided the organizational capacity to carry out an assessment that could link to the policy sphere as the IPCC links to the UN Framework Convention on Climate Change (Nilsson 2007).

Meanwhile, in the course of the mid to late 1990s, the need for better knowledge about climate change had emerged as a major issue also within the Arctic political cooperation, especially following the 1997/98 assessment from its largest working group, the Arctic Monitoring and Assessment Programme (AMAP). When IASC approached AMAP about an Arctic climate impact assessment in 1999, some planning for an assessment had already started and the time was ripe for a joint endeavor of IASC, AMAP, and one of the other Arctic Council working groups – Conservation of Arctic Flora and Fauna (CAFF); later labeled the Arctic Climate Impact Assessment (ACIA). After initial scoping, the ACIA was approved at the Arctic Council Barrow ministerial meeting in 2000 (Nilsson 2007).

The Barrow ministerial requested three reports: a scientific report, a popular science summary, and a policy document (Arctic Council 2000). Robert Corell was appointed as chair of the assessment steering committee. With the Arctic Council as the organizational setting for the assessment came a question of what knowledge was relevant to include as compared to IPCC's assessments, which are based almost exclusively on peer reviewed scientific literature. In the Arctic Council, indigenous peoples' organizations not only have a formal role as permanent participants; they also take part in scientific assessments. Because of this standing, combined with a push from indigenous peoples' representatives at key meetings and active interest from the ACIA chair to include indigenous knowledge, several chapters in the ACIA scientific assessment came to highlight indigenous observations and perspectives. This gave indigenous people a role as knowledge providers that they had not previously had in climate science. Moreover, and in contrast to the framing that emphasizes the Arctic as part of the global climate system, indigenous participation also brought emphasis to local impacts of climate change and how changes in the physical environment interact with social, cultural, and political processes (Nilsson 2007).

The ACIA highlights how new political circumstances in the Arctic – the Arctic Council as a regime – allowed space for indigenous peoples and created a new context for climate change science. If the IPCC brought scientists and nation states together as actors in framing climate change, the ACIA brought indigenous peoples as transnational non-state actors into the game. With this development came new notions about what is legitimate knowledge for policy making, with implications for future research in the Arctic including international polar years.

The challenge of western science as the only legitimate conveyer of policy relevant knowledge that emerged in the ACIA has also been visible within the IPCC. For example, the lack of legitimacy in the South forced the IPCC to both find ways to involve more scientists from the South and to address issues that are important to southern countries (Biermann 2006; Siebenhüner 2006). Southern scientists have also challenged how the IPCC initially framed key concerns in climate science. This included issues ranging from estimates of methane emission to highly value-laden questions about the economic value of human life.

The trend that has emerged is that knowledge production in the 1990s was increasingly intertwined with the global discourse on equity, an issue that surfaced already at the Stockholm Conference in 1972 and has since been on the agenda of international environmental politics (Linnér and Jacob 2005; Kjellén 2007). This trend is in turn closely intertwined with a critique against globalism, as defined by western science, as a sole legitimate framing of environmental problems. The development at the turn of the millennium can be placed in contrast to the promotion of scientific internationalism during the Cold War period some 50 years earlier. Moreover, equity issues in global politics have come to include knowledge production and framing of climate change. Science and environment are not longer politically neutral arenas. In the context of the IPCC, it has become important to find norms, such as geographical representation, to ensure the political legitimacy of the process without sacrificing credibility to the climate science community (Biermann 2006) and in the ACIA indigenous perspectives became integrated into the scientific assessment.

The debate about legitimate knowledge is not unique to climate science but actually more pronounced in relation to some other environmental issues. A case in point is biodiversity, where the inclusion of traditional knowledge is institutionalized at the global level and where scientific assessments highlight the meeting of knowledge traditions and scale perspectives (Capistrano et al. 2006; Reid et al. 2006). This discourse has also become pronounced in the emergence of science for sustainable development and the study of Earth systems governance, where issues such as access and allocation are high on the scientific agenda (Biermann 2007).

The parallel development in the Arctic is illustrated by the attitude towards indigenous peoples and their local traditional knowledge. In the AEPS and later Arctic Council, the predominant interests were initially formulated in national capitals south of the Arctic but have become increasingly moderated by concerns raised by Arctic indigenous peoples to the extent that attention to indigenous knowledge now is a strong norm within the Arctic Council. Nuttall described the situation as one in which indigenous environmental knowledge has been institutionalized and indigenous peoples perspectives were regarded almost uncritically as experts on environmental conservation (Nuttall 2000, 623; see also Tennberg 2000, 55 ff). In spite of the formal recognition of indigenous knowledge in the Arctic Council, there are also underlying controversies. A history of colonial knowledge production done without concern for people living in the Arctic raises issues about the ethics of research conduct and also who has the right to define the Artic environment (Tennberg 2000, 59: ICARP II Working Group 11 2005). Control over how the Arctic is framed by science is also part of indigenous peoples' asserting political independence where they have been able to use the environmental framing of the Arctic to create a role for themselves in the broad international context (Nuttall 2000, 624). A key example of this is the role that the image of Inuit interests played during the Stockholm Convention on Persistent Organic Pollutants (Downie and Fenge 2003) and more recently the their active role in raising awareness about climate change (see e.g. Watt-Cloutier et al. 2006). Although the challenge of western science's prerogative to define climate science is part of a more general trend, it appears that this is particularly visible in the Arctic, possibly because indigenous peoples are fairly well organized and also have the Arctic Council as a political forum to act in, in contrast to the IPCC and UNFCCC where nation states are the only legitimate political actors. It is certainly an important part of the political context for the emphasis on human dimensions in the 2007–08 IPY.

Discussion and conclusions

This account of the intertwined development of international society, climate change science, and polar research from the mid-1800s to the early years of this millennium provides a historical background for the attention to Arctic climate change and the focus on human dimensions during the IPY 2007–08. It highlights how networks for scientific cooperation develop and how they are shaped by a combination of changes in the structure and norms of international society and political ideologies over time.

At the time of the first IPY, climate science was linked to a wish to understand how ice had shaped the northern landscapes and to an emerging interest among meteorologists to build professional networks across national borders. Previously, climate science had been carried out in a nationalistic context, but helped along also by new communication technologies such as the railroad and the telegraph climate science now became part of actor networks that went beyond the nation state and national capitals as the major centers of calculation. The additional center was the International Meteorological Organization (IMO), which was created to coordinate meteorological measurements for weather forecasting and which also became the organizational home of the first IPY.

The initial networks of climate and polar science were fragile and had no backing by the organization of international political society. Yet, the fact that a second polar year was organized in 1932–33 shows that the seed of international cooperation in meteorology survived a strong re-emergence of nationalism during the war years. Within the scientific community, internationalism was in vogue, as evidenced by the growth of international scientific societies.

After the Second World War, the situation changed dramatically and the political setting for the IGY in 1957–58, was quite different from the first two polar years. The United States was investing heavily in research to further its military efforts during the Cold War and the funding also benefited climate change science. In addition, internationalism became a US political ideology in its wish to create a new world order. The reorganization of the IMO into the WMO as an agency under the UN in 1950, including the strong US initiative in this process, is an example of how it affected climate science. The WMO also exemplifies another development of importance for climate change research – the early stages of multilateral political cooperation and the fact that not only professionals and scientists but also states become actors in the development of meteorology. With new actors came new monetary and political resources that could strengthen the professional meteorological network. The IGY fit well into this political setting and further strengthened the meteorological network by far-reaching efforts to coordinate data collection. One of IGY's legacies is the international weather data centers that have later proven crucial in understanding how the climate is changing and for building computer models of future climate change. Another legacy of the IGY was the Antarctic Treaty and the creation of the SCAR as a network of polar scientists that later proved important also for Arctic climate science. A third legacy was the start of a cultural era where planet Earth as a whole was in focus.

The scientific and professional networks have since gained further organizational strength, for example when the Global Atmospheric Research Programme was launched in 1967 under the auspices of WMO and ICSU and when it was followed by the World Meteorological Research Programme in 1980. Together with other development such as growth of the environmental movement, political recognition of responsibility for transboundary air pollution at the Stockholm Conference in 1972, and increasingly close interactions between climate science and policy communities as exemplified by UNEP and the IPCC, the stage for the 2007–08 IPY is, yet again, quite different from the previous IPY/IGY efforts.

Another important development for the IPY has been the growth of a discourse on sustainable development that includes human dimensions of environmental issues in a way that was not as apparent during the IGY 50 years earlier. The focus on sustainable development and human dimension has also brought a new kind of politics to bear on climate science, including the IPY. This includes issues related to equity and representation in arenas of knowledge production as has been apparent at the global level in the IPCC. In polar research it has become especially pronounced in the Arctic where scientists and nation states are no longer the sole legitimate transnational actors and scientists no longer the only legitimate knowledge providers, but where political and scientific communities are accompanied by indigenous peoples. An example was the prominent inclusion of indigenous perspectives in the ACIA and how previously local knowledges of Arctic climate change have started to enter the international climate science-policy network. An analysis of the ACIA has shown that the global gaze from the IGY is now being complemented with bottom-up local gazes that place emphasis on new issues such as the complex interactions of climate, political, cultural, and social changes (Nilsson 2007).

In summary, the history of IPY/IGY mirrors changes in international society and the relationship between science and society. Today, climate science does not derive legitimacy only from quality in theory and methods but also from who is allowed to participate in knowledge making and ultimately from the same principles that underlie democratic political governance. The theme of human dimensions in the IPY 2007–08 can be seen as an expression of this shift. Another major change is the growth of a strong, politically supported international network that effectively links climate science and politics. Arctic climate change is indeed much more than a question of weather.

Acknowledgements

This chapter is based on work that was carried out for a PhD dissertation at the Department of water and environmental studies, Linköping University, Sweden, and with support from colleagues at the Centre for Climate Science and Policy Research.

References

Agrawala, S. (1998a), 'Context and Early Origin of the Intergovernmental Panel on Climate Change', *Climatic Change* 39:4, 605–20.
—— (1998b), 'Structural and Process History of the Intergovernmental Panel on Climatic Change', *Climate Change* 39:4, 621–42.
AHDR (ed.) (2004), *Arctic Human Development Report* (Akureyri: Stefansson Arctic Institute).
AMAP (1997), Arctic Pollution Issues: A State of the Arctic Environment Report (Oslo: Arctic Monitoring and Assessment Programme).
—— (1998), *AMAP Assessment Report. Arctic Pollution Issues* (Oslo: Arctic Monitoring and Assessment Programme).
Andersen, S.O. and Sarma, K.M. (2002), *Protecting the Ozone Layer. The United Nations History* (London: Earthscan).
Archer, C. and Scrivener, D. (2000), 'International Co-Operation in the Arctic Environment', in Nuttall and Callaghan (eds).
Arctic Council (2000), Arctic Council. Notes from the Second Ministerial Meeting. Barrow, Alaska, USA October 12–13, 2000. Including ACIA Implementation Plan Version 3.7.
Arrhenius, S. (1896), 'On the Influence of Carbonic Acid in the Air upon the Temperature of the Ground', *The London, Edinburgh, and Dublin Philosophical Magazine and Journal of Science* 41:237–76.

Barr, W. (1982), 'Geographical Aspects of the First International Polar Year, 1882–1883', *Annals of the Association of American Geographers* 73:4, 463–84.

Baylis, J. and Smith, S. (eds) (2001), The Globalization of World Politics. An Introduction to International Relations (Oxford: Oxford University Press).

Behr, S., Coen, R., Warnick, W.K., Wiggins, H. and York, A. (2007), 'IPY History Reflects Progress in Science and Society', *Witness the Arctic* 12, 1–4.

Biermann, F. (2006), 'Whose Experts? The Role of Geographic Representation in Global Environmental Assessments', in Mitchell et al. (eds).

—— (2007), '"Earth System Governance" as a Crosscutting Theme of Global Change Research', *Global Environmental Change* 17:3–4, 326–37.

Bolin, B. (2007), The History of the Science and Politics of Climate Change. The Role of the Intergovernmental Panel on Climate Change (Cambridge: Cambridge University Press).

——, Döös, B., Jäger, J. and Warrik, R.A. (eds) (1986), *The Greenhouse Effect. Climate Change and Ecosystems* (Chichester: John Wiley and Sons).

Bravo, M.T. and Sörlin, S. (eds) (2002), *Narrating the Arctic. A Cultural History of Nordic Scientific Practice* (Canton, MA: Science History Publications).

—— —— (2002), 'Narrative and Practice – an Introduction', in Bravo and Sörlin (eds).

Bryld, M. and Lykke, N. (2000), Cosmodolphins. Feminist Cultural Studies of Technology, Animals and the Sacred (London, New York: Zed Books).

Bull, H. (1977), The Anarchical Society. A Study of Order in World Politics (New York: Palgrave).

Capistrano, D., Samper, C.K., Lee, M.J. and Raudsepp-Hearne, C. (eds) (2006), *Ecosystems and Human Well-Being. Multiscale Assessments Vol. 4* (Washington, Covelo, London: Island Press).

Clark, W.C., Jäger, J., Cavender-Bares, J. and Dickson, N.M. (2001), 'Acid Rain, Ozone Depletion, and Climate Change: An Historical Overview', in Social Learning Group (ed.).

Crawford, E. (1992), Nationalism and Internationalism in Science, 1880–1939. Four Studies of the Nobel Population (Cambridge, MA: Cambridge University Press).

—— (1998), 'Arrhenius' 1896 Model of the Greenhouse Effects in Context', in Rodhe and Charlson (eds).

Doel, R.E. (2003), 'Constituting the Postwar Earth Sciences: The Military's Influence on the Environmental Sciences in the USA after 1945', *Social Studies of Science* 33:5, 635–66.

Downie, D.L. and Fenge, T. (eds) (2003), *Northern Lights Against POPs* (Montreal and Kingston: McGill–Queen's University Press).

Dunne, T. and Schmidt, B. C. (2001), 'Realism', in Baylis and Smith (eds).

Edwards, P.N. (2001), 'Representing the Global Atmosphere', in Miller and Edwards (eds).

—— (2004), '"A Vast Machine": Standards as Social Technology', *Science* 304, 827–28.

Ekström, A. (ed.) (2004), *Den mediala vetenskapen* (Nora: Nya Doxa).

Farrell, A.E. and J. Jäger, J. (eds) (2006), Assessments of Regional and Global Environmental Risks. Designing Processes for the Effective Use of Science in Decisionmaking (Washington, DC: Resources for the Future).

Frängsmyr, T. (1982), 'Polarforskning – Från Hjälte Till Vetenskapsman', in Swedish Royal Academy of Sciences (ed.).

Franz, W.E. (1997), 'The Development of an International Agenda for Climate Change: Connecting Science to Policy', *ENRP Discussion Paper E-97-07,-33*. Boston: Kennedy School of Government, Harvard University.

Greenaway, F. (1996), Science International. A History of the International Council of Scientific Unions (Cambridge: Cambridge University Press).

Gupta, J., van der Leew, K. and de Moel, H. (2007), 'Climate Change: a "Glocal" Problem Requiring "Glocal" Action', *Environmental Sciences* 43, 139–48.

Hart, D.M. and Victor, D.G. (1993), 'Scientific Elites and the Making of US Policy for Climate Change Research 1957–74', *Social Studies of Science* 23:4, 643–80.

Heininen, L. (2004), 'Circumpolar International Relations and Geopolitics' in AHDR (ed.).

Hild, C. (2004), 'Human Health and Well-Being', in AHDR (ed.).

Holmberg, G. (2004), 'Nils Ekholm, Stormvarningarna Och Allmänheten', in Ekström (ed.).

Houghton, T., Jenkins G.J. and Ephraums, J.J. (eds) (1990), *Scientific Assessment of Climate Change – Report of Working Group I* (Cambridge: Cambridge University Press).

ICARP II Working Group 11 (2005), *ICARP II. Working Group 11. Arctic Science in the Public Interest. Science Plan* <www.arcticportal.org/iasc/icarp>, accessed 26 August 2008.

IPCC (1997), 'The Regional Impacts of Climate Change. An Assessment of Vulnerability. Summary for Policymakers. Special Report of Working Group II'. Wienna: IPCC Secretariat <www.ipcc.ch>.

IPY 2007–2008 (2005), 'History of IPY' <http://www.ipy.org/development/history.htm>, accessed 13 April 2007.

Jasanoff, S. (2004a), 'Heaven and Earth: The Politics of Environmental Images', in Jasanoff and Long Martello (eds).

—— (2004b), 'Ordering Knowledge, Ordering Society' in Jasanoff (ed.).

—— (ed.) (2004c), States of Knowledge. The Co-production of Science and Social Order (London and New York: Routledge).

—— and Wynne, B. (1998), 'Science and Decision Making', in Rayner and Malone (eds).

—— and Long Martello, M. (eds) (2004), *Earthly Politics. Local and Global in Environmental Governance* (Cambridge, MA: MIT Press).

Johnson Theutenberg, B. (1982), 'Polarområdena – Politik Och Folkrätt', in Swedish Royal Academy of Sciences (ed.).

Kellogg, W.W. (1987), 'Mankind's Impact on Climate: The Evolution of Awareness', *Climatic Change* 10:2, 113–36.

Keohane, R.O. and Nye, J.S. (1994), *Power and Interdependence* (Harper Collins).

Keskitalo, C. (2002), 'Region-Building in the Arctic: Inefficient Institutionalism? A Critical Perspective on International Region-Building in the "Arctic"', ISA Annual Convention, March 24–27, 2002, ISANET <http://www.isanet.org/noarchive/keskitalo.html>, accessed 23 October 2003.

Kjellén, B. (2007), *A New Diplomacy for Sustainable Development* (New York London: Routledge).

Latour, B. (1987), *Science in Action* (Cambridge, MA: Harvard University Press).

Lear, L. (1997), *Rachel Carson. Witness for Nature* (New York: Henry Holt and Company).

Liljequist, G. (1982), 'Det Internationella Geofysiska Året 1957–58 – Den Moderna Polarforskningens Startår', in Swedish Royal Academy of Sciences (ed.).

Linnér, B.-O. (2003), The Return of Malthus. Environmentalism and Post-War Population-Resource Crisis (Isle of Harris: White Horse Press).

—— and Jacob, M. (2005), 'From Stockholm to Kyoto and Beyond: A Review of the Globalisation of Global Warming Policy and North–South Relations', *Globalization* 2:3, 403–15.

Litfin, K.T. (1994), Ozone Discourses. Science and Politics in Global Environmental Cooperation (New York: Colombia University Press).

Lundgren, L.J. (1997), *Acid Rain on the Agenda* (Lund: Lund University Press).

Miller, C.A. (2001), 'Scientific Internationalism in American Foreign Policy: The Case of Meteorology, 1947–1958', in Miller and Edwards (eds).

—— and Edwards, P. N. (eds) (2001), *Changing the Atmosphere. Expert Knowledge and Environmental Governance* (Cambridge, MA: MIT Press).

—— —— (2001), 'Introduction', in Miller and Edwards (eds).

Mitchell, R.B. Clark, W.C., Cash, D.W. and Dickson, N.M. (eds) (2006), *Global Environmental Assessments: Information and Influenc* (Cambridge, MA: MIT Press).

Morgenthau, H.J. and Thompson, K.W. (1993), *Politics Among Nations. The Struggle for Power and Peace* (New York: McGraw-Hill).

Nationalencyclopedin (1994), 'Meteorologi' (Stockholm: Bra Böcker).

Nilsson, A.E. (2007), *A Changing Arctic Climate. Science and Policy in the Arctic Climate Impact Assessment*. Dissertation, Dep. of Water and Environmental Studies, Linköping University (Linköping University Press: <http://urn.kb.se/resolve?urn=urn:nbn:se:liu:diva-8517>).

Nuttall, M. (2000), 'Indigenous Peoples' Organisations and Arctic Environmental Cooperation', in Nuttall and Callaghan (eds).

—— and Callaghan, T. V. (eds) (2000), *The Arctic. Environment, People, Policy* (Amsterdam: Harwood Academic Publishers).

Rayner, S. and Malone, E.L. (eds) (1998), *Human Choice and Climate. Vol. 1. The Societal Framework* (Columbus, OH: Battelle Press).

Reid, W.V., Berkes, F., Wilbanks, T.J. and Capistrano, D. (eds) (2006), *Bridging Scales and Knowledge Systems. Concepts and Applications in Ecosystem Management* (Washington: Island Press).

Revelle, R. And Seuss, H.E. (1957), 'Carbon Dioxide Exchange Between the Atmosphere and Ocean and the Question of an Increase of Atmospheric CO_2 During the Past Decades', *Tellus* 9, 18–27.

Rodhe, H. and Charlson, R. (eds), *The Legacy of Svante Arrhenius Understanding of the Greenhouse Effect* (Stockholm: Royal Academy of Sciences and Stockholm University).

Rosentrater, L. (ed.) (2006), 2° is too much! Evidence and Implications of Dangerous Climate Change in the Arctic (Oslo: WWF).

Schram Stokke, O. (1990), 'The Northern Environment: Is Cooperation Coming?' *Annals of the American Academy of Political and Social Science* 512:Nov., 58–68.

Selin, H. and Linnér, B.-O. (2005), *The Quest for Global Sustainability: International Efforts on Linking Environment and Development. CID Graduate Student and Postdoctoral Fellow Working Paper No 5* (Boston: The Center for International Development at Harvard University).

Serafin, R. (1996), 'How Did Our Weather Measurement Systems Evolve?' *Science Now* 3:2, <http://www.proquestk12.com/curr/snow/snow596/snow596.htm# weather>, accessed 26 August 2008.

Siebenhüner, B. (2006), 'Can Assessments Learn, and If So, How? A Study of the IPCC', in Farrell and Jäger (eds).

Social Learning Group (ed.) (2001), *Learning to Manage Global Environmental Risk. Vol. 1* (Cambridge, MA: MIT Press).

Steffen, W., Jäger, J., Carson, D.J. and Bradshaw, C. (eds) (2002), *Challenges of a Changing Earth* (Berlin: Springer).

Swedish Royal Academy of Sciences (ed.) (1982), *Polarforskning. Förr, nu och i framtiden* (Stockholm: Swedish Royal Academy of Sciences).

Tegart, M.J.M., Sheldon, G.W. and Griffiths, D.C. (eds) (1990), *Impacts Assessment of Climate Change – Report of Working Group II* (Australian Government Publishing Service).

Tennberg, M. (2000), *Arctic Environmental Cooperation: A Study in Governmentality* (Hants and Burlington: Ashgate Publishing Company).

Waltz, K.N. (1979), *Theory of International Politics* (New York: McGraw-Hill).

Watt-Cloutier, S., Fenge, T. and Crowley, P. (2006), 'Responding to Global Climate Change: The View of the Inuit Circumpolar Conference on the Arctic Climate Impact Assessment', in Rosentrater (ed.).

Weart, S.R. (2003), *The Discovery of Global Warming* (Cambridge, MA: Harvard University Press).

—— (2004), 'The Discovery of Global Warming', www.aip.org/history/climate, accessed 26 August 2008.

Weatherhead, E.C. (1998), 'Climate Change, Ozone, and Ultraviolet Radiation', in AMAP (1998).

WMO (2005a), 'IMO: The Origin of WMO' <www.wmo.ch/wmo50/e/wmo/history_pages/origin_e.html>, accessed 18 October 2005.

—— (2005b), 'The Beginnings of WMO (1950s–1960s)' <www.wmo.ch/wmo50/e/history_e.html>, accessed 18 October 2005.

—— (2005c), 'The Historical Roots of WMO' <www.wmo.ch/wmo50/e/history_e.html>, accessed 18 October 2005.

Worster, D. (1994), *Nature's Economy. A History of Ecological Ideas* (Cambridge: Cambridge University Press).

Wråkberg, U. (2006), 'Nature Conservationism and the Arctic Common of Spitsbergen 1900–1920', *Acta Borealia* 23:1, 1–23.

Young, O.R. (1985), 'The Age of the Arctic', *Foreign Policy* 61, 160–79.

—— (1998), *Creating Regimes: Arctic Accords and International Governance* (Ithaca, NY: Cornell University Press).

—— (2002), The Institutional Dimensions of Environmental Change. Fit, Interplay, and Scale (Cambridge, MA: MIT Press).

—— and Osherenko, G. (eds) (1993), *Polar Politics. Creating International Environmental Regimes* (Ithaca and London: Cornell University Press).

—— and Einarson, N. (2004), 'A Human Development Agenda for the Arctic: Major Findings and Emerging Issues', in AHDR (ed.).

——, King, L.A. and Schroeder, H. (eds) (2008), Institutions and Environmental Change: Principal Findings, Applications, and Research Frontiers (Cambridge, MA: MIT Press).

Chapter 2

Revisiting Politics and Science in the Poles: IPY and the Governance of Science in Post-Westphalia

Jessica M. Shadian

We cannot study and understand changes in the way science is produced independently from changes taking place in society at large (Pestre 2003, 247).

Introduction

Over the course of the last several decades, it has become widely recognized that global politics is in a period of transformation. The role and authority of the inter-state system as a primary facilitator of knowledge is undergoing considerable change and scholars have begun to question the ongoing assumptions of Westphalian politics as traditionally understood (Ruggie 1993; Rudolph 2005; Osiander 2001; Krasner 1993, 1995–1996, 1999, 2001; Agnew 2005). Historically, the examination of states has been the central institutional means by which to assess and address the direction of global political change. Yet, with the emergence of a host of new non-state actors as formal participants in global politics, this traditional focus for analyzing the ways in which knowledge is constructed and governed has shifted to include a broad range of new and emerging institutions. Polar science programmes generally are one tradition that has received limited examination as a sphere in which to examine these shifting contours of global politics. The history of the IPY, in particular is one such polar programme which has the potential to help illuminate better understandings of the ongoing and contextual relationship between changing assumptions of science and governance as new non-state actors increasingly come to define and determine the way in which the governance of Arctic science is played out.

Rather than searching for historical facts, this chapter offers one particular narrative of the IPY as an ongoing history co-constituted by the history of global politics and particularly modern Westphalia. According to historian Geoffrey Roberts, 'the ontology of the human world is defined by the existence and ubiquity of private and public narratives (including, of course, the stories of historians) ... narrative historians tell stories about the world because that is the way they see their world' (Roberts 2006, 710; also see Jackson 2008). This narrative, viewing the IPY as an ongoing process of scientific thought and political change, is broken down into three sections. The first offers a brief discussion of theory as it relates to recent changes in global politics – namely the shift from international relations (Westphalia) to global governance (post-Westphalia). The second is divided into two parts that include an abridged narrative of the political history of modern Westphalia and likewise a brief history of the IPYs. The importance of these narratives is not to re-invent the impacts of enlightenment thought, exploration, colonialism, territorial and natural resource development, but rather to question how and where these legacies persist today and how they intersect with the changing basis upon which global politics, law and scientific inquiry proceeds.

Finally, this chapter provides three snapshots of the fourth IPY that highlight political and scientific legacies and change. Focusing on the role of new non-state actors, the first instance looks at the shifting relationship between international law, politics, and climate change science as played out through the fourth IPY. This case study perceives the science community as a norm entrepreneur in shifting the basis upon which international law has traditionally proceeded (state rights) to individual and collective rights and responsibilities. The second case concerns the role of indigenous politics as a new formal actor in the IPY and the ways in which indigenous agency is redefining the boundaries and meaning of Arctic science. Lastly, the third example points to the role of private industry as another example of new non-state actors defining the role and agenda of IPY science. Focusing on the expanding range of political actors who help to construct and define the political and scientific spaces of the Arctic, the narrative put forth in this chapter draws out the ongoing constitutive relationship between science and politics and the dynamic ways in which IPY science is defined and constructs new and lasting sites of authority over knowledge. Through a historical and political narrative of IPY science, this chapter becomes one piece, in fact, of a larger narrative aimed to better understand and reflect on the processes of global political change including who has sovereignty over scientific knowledge and how the IPY reflects shifting power and authority of this knowledge over time.

The politics of knowledge: A view of science to proceed by

> In a world where knowledge is power, one should thus expect that the activities
> of science are largely under the control and direction of those sectors of society
> that hold dominant political power (Schwartz 1996, 148).

The ontological underpinnings of IPY science cannot be abstracted from larger intellectual modes of thought. The first IPY (1882–83) was established with the intention of expanding our understandings of some of the most intricate scientific puzzles of the time; particularly the scientific gaps in meteorology. Since then, the intellectual landscape has markedly transformed. In particular, the recognition and acknowledgment of human-induced environmental change has unseated the ongoing assumptions of enlightenment rationalism separating humans from the environment. The decreasing rather than increasing understanding – much less ability to control – weather patterns and lingering effects of human induced environmental catastrophes (i.e. Chernobyl, Bhopal, and Exxon) have helped call into question previously held assumptions that with the right variables and enough data we can learn the truth about, and therefore control, the natural world.

Environmental issues alongside a changing political landscape – including decolonization in many parts of the world – have helped shift the meanings and practices of science. In particular, this latest intellectual inquiry has shifted the frontiers of scientific thought from an ahistorical enterprise into that which is historically dependent and, in part, socially constructed. This 'discovery' in the natural sciences has been compounded by a broader cultural turn in the social sciences (Jackson 2008). This shift highlights both the relationship between society and nature and further the co-constitution between the social world and the object of study. According to Laclau, 'we are always internal to a world of signifying practices and objects' (Howarth et al. 2000, 3). Science and social thought, according to Milton Freeman are part of an interdependent, dynamic and complex set of 'system relationships' (Freeman 1992) not only within nature and society but also between nature and society. As such, scientists have begun to question the desire for causality and traditional beliefs that isolating variables and collecting data will lead to the ability for prediction. Rather, attention has turned to the constitutive relationships (or processes) as they contain the ability to account for and analyze the ongoing changing conditions between and within the natural and social world.

In the social sciences in particular, this transformation in intellectual thought is considered part of a larger epistemological shift or a new paradigm, referred to by some as the 'knowledge society' (Nowotny et al. 2003, 185). This shift has been accompanied by the concession of the co-constitution between intellectual thought and political institutions. Specifically, the end of the Cold War and onset of globalization exacerbated the fallacy that has until recently underpinned the foundation of international politics. States, rather than conceived as atomic individual and rational entities floating about in an international system, are not only a constitutive facet of this system (able to endure only as long as the global political system endures) but likewise cannot remain isolated from the politics of others. Increasingly this has included 'others' which are not in the form of states. Whereas traditional Westphalian politics was a politics of interaction among sovereign states, in a post-Westphalian political world territorial borders (a static notion) remain; yet goods, people, transboundary environmental issues and knowledge have proven to be dynamic and complex. Moreover, the conviction that knowledge is socially constructed rather than having a predetermined order has provided the means by which to perceive science as cultural rather than universal and static. As such, to build a narrative focusing on the politics of science requires an ontological assumption that the construction of knowledge and politics are mutually constitutive and it is within this process that power is created, sedimented and broken down.

Science and development

Focusing on processes opens a space to examine contending perspectives of knowledge and accordingly the politics of development. The relationship between science and politics is most evident in changing policies and legal sentiments of international development, particularly through the discourse of sustainable development (brought into the international forum with the Brundtland Report). More recent attention to climate change has augmented this prevailing ideology while also adding a new underlying assumption regarding human agency. Rather than seeking to control the environment in a sustained manner the discourse surrounding climate change is that of adaptation and resilience. As such nature itself has attained a sense of agency (Latour 1999) and the scientific aim has become to find the means and knowledge for humans to endure and adapt in a changing physical world. According to Arctic researchers, resilience and adaptation is defined as 'proactive management of change to foster both resilience (sustaining those attributes that are important to society in the face of change) and adaptation (developing new socioecological configurations that function effectively under new conditions)' (Chapin et al. 2006). Climate change science most astutely highlights the centrality placed on recognizing the relationship and shifting centres of agency between the social and natural world. Climate change, according to Eugene Skolinkoff, 'is the apotheosis of the idea that "everything is related to everything else"' (Skolinkoff 1994, 183). As studying weather is one of the primary foundations of the IPY tradition (see Nilsson in this volume), from a historical perspective, this continuation through a focus on climate change opens a space to examine the IPY as part of a larger narrative of global political, legal and social change.

Locating spaces of change: From Sputnik to global climate change

The history of IPY science cannot be abstracted from the particular global circumstances in which each IPY has taken place. These historical eras have helped structure the aims and focus of scientific inquiries taking place during each IPY (1882–83; 1932–33; 1957–58; 2007–08). The last IPY – the IGY – for instance, materialized in the midst of the Cold War. Given its timing, it could be argued that IGY was intrinsically part of Cold War technological advancement while simultaneously serving as a harbinger of the intellectual and political changes following shortly thereafter.

In May 1954, the Soviet Union embarked on a project to develop the first Soviet intercontinental ballistic missile (NASA 2008). The interest leading up to this decision was largely based on the work by the Soviet researcher Sergei Pavolvich Korolev, coupled with written statements from the US regarding interest in satellite launches. The growing international interest in science satellites eventuated in a US-led resolution passed in October 1954 by the IGY governing board calling for science satellite launches during the IGY. The Rome resolution, as it came to be called, 'pulled back the hammer on the starter's gun in the satellite race (NASA 2008).' When the IGY commenced in July 1957, only two months later: 'Six minutes after liftoff', PS-1 – soon to be renamed Sputnik – ejected from its expended carrier rocket to became [*sic*] a second moon of Earth. A new Age of exploration was under way (NASA 2008).

The international scientific cooperation produced by the IGY and underpinning Sputnik, however, was soon overshadowed by escalating political tensions (Sagdeev and Logsdon 2008). Nevertheless, given the international political circumstances in which the IGY took place, the Cold War helped bring attention to and boost the perceptual importance of the IGY as Big Science at the international level – an ongoing legacy which is continuously commemorated to this day in varying capacities (e.g. see NOAA 2007).

The legacies of the IGY, however, extend well beyond symbolic commemoration. The overall international political mood following the end of the IGY soon gave way to a new global fixation. In December 1968, Apollo 8 was launched by the US in reaction to Sputnik. While Apollo 8 was considered an affirmation of US Cold War strength, it also sent back live photos of the earth from space and with it produced – for the first time – a sense of 'our' planet; an event which has since been equated with the birth of a new environmental consciousness. Apollo reconceptualized the earth as finite, fragile and demanding protection from the impacts of human industrial development. National and international policies to protect the Earth soon emerged. This growing perceptual shift from a Cold War world to an environmentally precarious world eventually made its way to the Arctic. The end of the Cold War in particular, Carina Keskitalo contends, activated an international transformation of the Arctic from being either entirely ignored or considered uninhabited into an 'international region conceived primarily on the basis of environmental and indigenous concerns' (Keskitalo 2002, 2). According to Keskitalo, the 'Arctic' has come to be perceived as a frontier by those who seldom inhabit the region themselves but consider it a potential for their society's economic growth. This predominately North American 'descriptional frame of the Arctic' posited the North as a wildland in the eyes of environmentalists yet also a homeland to its indigenous people (Keskitalo 2002, 6).

In the European and Russian Arctic, a focus on energy and other natural resource development existing in tandem with the increasing political agency of Europe's Arctic indigenous peoples have created a contending Arctic discourse. Rather than a wild frontier to be preserved, European and Russian Arctic political agendas (i.e. Barents region) constructed a post–Cold War Arctic where resource development and sustainable resource management were up front and centre.

Putting these emerging Arctic narratives into practice, the Arctic Council was founded in 1996. The inauguration of the Arctic Council paralleled emerging circumstances and dominant global trends. This includes, in tandem with the growing global environmental consciousness, the further emergence of multispacial governing institutions as well as the growing power and influence of a host of new non-state actors (Young 1999; Betsill and Bulkeley 2004, 471–93; Bulkeley 2005, 875–902). For instance, the Arctic Council is the only pan-Arctic regime and is comprised of the eight Arctic states and six indigenous organizations: Arctic Athabaskan Council (AAC), Gwich'in Council International (GCI), Inuit Circumpolar Council (ICC), Russian Association of the Indigenous Peoples of the North (RAIPON), and Sámi Council (SC) and the Aleut International Association (AIA). Within this new political milieu of the Arctic Council, non-state actors are no longer merely considered exogenous or even endogenous epistemic communities. Social scientists now debate over how to understand and to what extent new non-state actors – including NGOs, local governments, transnational corporations, indigenous and other ethnic actors – play in a deterritorializing and changing global political arena. Rather than providing supplemental insights, non-state institutions (whether sub-state, transnational or international) are part the foundation upon which dialogue, governance and the construction of knowledge proceeds. These changes, according to Peter Haas, are emblematic of a move towards global non-polarity in which: 'there are many more power centers, and quite a few of these poles are not nation-states. Indeed, one of the cardinal features of the contemporary international system is that nation-states have lost their monopoly on power and in some domains their preeminence as well. States are being challenged from above, by regional and global organizations; from below, by militias; and from the side, by a variety of nongovernmental organizations (NGOs) and corporations. Power is now found in many hands and in many places (Haas 2008)'. It is within this political landscape that, the timely occurrence of the fourth IPY serves as a valiant instance of post-Westphalian or, in Haas' terms, non-polar politics.

The world according to Westphalia: Constructing sovereign assumptions

Science, politics, economics and society are historically dynamic and constitutive processes of a larger narrative about nation state-building and the construction of the Westphalian political system. These interconnected processes have continuously defined and redefined local, state, regional and international identities. Moreover, they have structured and been structured by changing intellectual modes of thought and definitions of progress and development. Within this tradition, scientific exploration has been one of the hallmarks of securing and expanding sovereign authority over territory.

The 1649 peace of Westphalia brought an end to the German 30 Years War and an 80-year war between Spain and the Netherlands. The accompanying Treaties of Munster and Osnabruck initiated the making and evolution of the modern international state system. This historical story of Westphalian international relations can be regarded as a chronicle of shifting sovereignty, resources, and territorial ownership. Westphalia put into place a system for dividing authority, where sovereignty was symbolized through the political affirmation and ultimately dominance of the nation-state. Westphalia, as such, constructed a new set of inside-outside boundaries reconstituting the limits and domain of sovereign political space, a space which has continuously been contested and reshaped over time.

In the world according to Westphalia, necessary for any polity to be considered sovereign was the possession of a bounded territory by which to build its narrative (Agnew 2005; Anderson 1991; Ansell and Weber January 1999; Bhabha 1990; Osiander 2001; Rudolph 2005). The Westphalian narrative comprises stories of territorial expansion, colonial conquest, and the search for new economies beyond domestic frontiers; efforts which have become a central facet in the construction of individual national histories. By the eighteenth century, territory in international relations permeated all political endeavours and symbolized all formal power. Into the nineteenth century territorial borders had constructed the global map into either sovereign states or colonial territories and the state system gradually took on an assumed status – a sedimented belief in a shared history of an international system. It was this *a*historical essence – a belief that it had always been this way – which modern international relations theory would eventually be founded upon. Incidentally, sovereignty became an assumed prerogative of the state marked by a particular territory. Through this historical progression, collective polities without a state were made to exist separate from the accompanying national narratives. As colonial conquest and territorial expansion often became a further means for strengthening the nation-state internationally (as territory represented power), indigenous peoples, while physically left outside of this nation-building narrative, were subsumed similar to natural resources in need of development thereby becoming part of the national narrative of the sovereign state. Global politics as such became the politics of states.

Westphalian sovereignty over time evolved into two legal stratums, the domestic and the international. Internationally, Westphalian sovereignty is marked by the explicit right to control all things within defined state boundaries, the assurance of non-intervention and an overall indifference to the domestic circumstances of other states. Domestic sovereignty, based on land ownership, denoted total control over the means of resource and economic development, which was then legitimated through international law (Held and McGrew 2003). The ability for a state to appropriate land considered *terra nullius* is the international legal doctrine of discovery and occupation. Most often discovery is not considered an adequate means in and of itself, but further, according to Judge Huber, requires an 'effective and continuous display of territorial sovereignty ... [of] not merely of sighting but also ... some form of symbolic annexation or act of taking possession' (For more on Judge Huber, see Island of Palmas Case of 1928 in Jacobsson 2004). Consequently, territorial sovereignty has historically gone hand in hand with scientific exploration and research effectively serving as natural extensions of state ambitions in claiming specific territories (i.e. cartography demarking land, the construction of field stations, fisheries, mining and other sciences related to the exploitation and development of natural resources).

The Arctic, far from immune to these events, became a pivotal piece of Western scientific exploration and colonial conquest. The popular tales of polar exploration did not only make national heroes but were equally considered scientific endeavours in their own right, aiming to uncover the mysteries of the polar regions and its peoples. Conquering the Arctic, as such, was perceived as the height of human endeavour – the Arctic as the last frontier. Subsequently, embedded within this historical narrative of Westphalia is also the story of Arctic colonialism. Westphalia put into place new centres of authority and new boundaries of inclusion and exclusion between European 'discoverers' and the Arctic's indigenous peoples. When European explorers arrived in Northern North America, for instance, they found that no one *owned* the land. Subsequently, the indigenous peoples living throughout the region were appropriated by governments elsewhere alongside that of the land and its resources.

Conquering land and resources was driven by another shared understanding, which assumed that nature existed for human consumption and economic gain. The industrial revolution was predicated on and driven by the strengthening prevalence of enlightenment thought, which created new means while also structuring the ways in which science, economics, politics, and society interacted and unfolded. By the end of the industrial revolution enlightenment thought had acquired an ahistorical aura of its own permeating all social and physical scientific inquiry.

Undercurrents to restructuring the traditional foundations of the Westphalian political system swelled during the 1960s, with the emergence of the environmental movement and as much of the colonized world (e.g. Asia and Africa) began to assert rights to self-determination. The objective of self-determination during the 1960s – statehood – was based on territorial integrity. This was the realization of the existing political framework where sovereignty meant the combination of territory, and national identity was realized in the form of an independent state (Rudolph 2005, 7). Yet, by the 1970s – when the Arctic's indigenous peoples also began to assert their own rights in global political processes – the international architecture within which those rights could be asserted was already in a process of reconstruction. The politics of the Arctic's indigenous peoples, while differing from the external colonization in other parts of the world, was similar in that it was nevertheless colonization within already existing states.

Arctic indigenous aims for self-determination from the 1970s (though the Sámi Parliament was established in 1956) and onwards have taken place in concert with a larger process unbundling traditional assumptions of state sovereignty in international politics. Often, rather than aiming for state sovereignty or territorial integrity, Arctic indigenous peoples seek political agency through the 'right' to develop their own land and resources otherwise referred to as *cultural integrity* (which includes validation of traditional indigenous scientific practice for determining the course of development alongside political and economic rights). The indigenous land claims agreements which unfolded at this time and thereby formal incorporation of indigenous peoples into the narrative of Arctic and global politics have been central to the construction of an Arctic discourse focused on sustainable development and accompanying science policies and practices. Evidence of this shift is exemplified in the mandates and projects of this fourth IPY in which indigenous participation and knowledge are central to the practices of science in the Arctic. Indigenous 'rights' regarding the politics of Arctic science speak to the larger global trends regarding increased non-state political agency. As new polities are increasingly able to stake a claim in global politics, indigenous agency is only one new stakeholder with the capacity to help steer and guide Arctic and global political processes.

A Westphalian narrative of the International Polar Year (IPY)

> Each nation published their observations independently and the International Polar Commission subsequently dissolved. … 'It may be that if the publication, and above all the discussion of the observations had been left to a central office, possibly international, the scientific level of the work accomplished would have been better appreciated.' Arktowski (1931) in NOAA Arctic Research Office project on first IPY (NOAA Arctic Research Office)

The IPY as an international science tradition was as much a part of the Westphalian narrative as the age of discovery or the Cold War. Historically, the IPY proceeded on the assumption that international coordination is necessary in order to enhance scientific understandings. The scientific aim, however, has also been driven by the belief that the results of IPY science would directly contribute to nation-building aims at the individual state level. The concept of the IPY was first proposed by the Austrian Naval lieutenant, Karl Weyprecht, during the first International Meteorological Congress in Vienna in 1873. Weyprecht believed that 'if only polar expeditions could be simultaneous and make specific observations using standardized apparatus at different geographical points, then the results would be more valuable and … many of the mysteries of nature would be solved more readily …' (Martin 1958, 19). Central to his initiative was a conviction that geographical exploration and scientific investigation should be linked (Martin 1958, 19). The results of the IPY would then to be exchanged through publication (Luedecke 2004).

Fifty years following the first IPY, a second IPY was launched and this time the science also included cosmic radiation and ionosphere (Luedecke 2004, 20). It also maintained its focus on observation and the exchange of results thereby creating a tradition of international polar science collaboration. A central aim of the first two IPYs was to produce long-term isolated data which could help predict weather patterns. Weather prediction would provide better sea navigation and improved communications, both of which were central to discovery, colonization, and resource exploitation. While the Arctic's indigenous peoples made a distinct contribution to the scientific endeavours taking place in Arctic field stations (Sorlin 2007 and see Wrakbeg in this volume), their involvement consisted, at best, of casual labour rather than being considered as partners in science (much less scientists in their own right). The Inuit, for example, have been recognized in historical accounts by researchers as providing knowledge needed for survival in the Arctic. However, this indigenous knowledge was not regarded as 'science' and thus left outside of the formal scientific narratives (Brody 1990 and 2002; Stefansson 1943; Beach 2001).

Since the second IPY, new tools of science have emerged. According to Dominique Pestre, a new turn towards the technosciences can be found to exist in the early part of the twentieth century resulting in a transformation of the field that reached its height during the Cold War. Throughout the eighteenth and nineteenth century in general, Pestre asserts, nation-building included the nationalism of science. Yet, despite this general trend, it was during the Cold War that the link between science and the state reached its apex (Pestre 2003, 250). The Cold War led to the introduction of the industrial research laboratory as well as heightening the link between universities, state-systems, the military, business, and social interests (Pestre 2003, 249).

The proposal for the third IPY was approved by the International Council of Scientific Unions (ICSU) in October 1951 and was inspired by a small group of scientists who met in Maryland, USA. The intent of the third IPY was to introduce new means of observation including cosmic rays, geodetic observations of the moon, and rockets (Chapman 1954, 924). Accordingly it was determined that the third IPY would include the study of all altitudes and depths. The earth's atmosphere, surface, and interior were chosen as the focus 'so as to obtain as complete a picture as possible of the physical forces acting on the whole earth' and the title, International Geophysical Year (IGY) conferred to reflect that emphasis (Martin 1958, 20). At the height of the Cold War, the IGY, while expected to be 'written into the history books because it recounts how humankind made its break through to the study of space beyond the atmosphere of his own planet earth' (Martin 1958, 18), was moreover driven by desires for greater state power. And ironically, the scientific aims to preserve state security dominating the IGY – while encouraging new ways to understand the earth from space – also drew attention to the global environmental degradation that human industrial development had created through the course of recent history in the drive for state power (and greatly amplified by the Cold War). The Arctic, the very point where the East met the West, was one of the largest recipients of nuclear weapons stockpiles, military radar stations, nuclear testing and human displacement (Jacobsson 2007, 307).

The aims of the fourth IPY were once again to understand weather patterns and the earth's atmosphere. While the scientific aims of the fourth IPY remained consistent with its predecessors, the assumptions for the need to understand weather change have significantly transformed since the first IPY. In particular, rather than solely aimed to understand the earth's natural environment, the fourth IPY also focused on the human aspects of weather change and incorporated the social sciences into its research mandate. This included specific attention to not only the livelihood and scientific knowledge of the indigenous peoples who inhabit the Arctic, but also the formal recognition that indigenous knowledge is central for understanding and assessing the state of the natural environment and attaining sustainable development policy and practice.

As such, the official aims of the fourth IPY was to better understand physical and social changes in the polar regions. This included enhancing observatories, investigating the frontier of science in the polar regions, as well as the 'cultural, historical, and social processes that shape the sustainability of circumpolar human societies, and to identify their unique contributions to global culture diversity and citizenship' (Jacobsson 2007). The direct focus on the societal connection to the natural environment of the fourth IPY is considered by many to be a unique turn in the tradition of the IPY.

It takes into account for the first time the social sciences (Though Krupnik et al. (2005) has pointed to a long standing legacy of social science participation in the history of IPY) and further, formally includes the contribution of indigenous knowledge for understanding the global state of the environment. On the surface the aims of the fourth IPY may seem to signal a departure from the IPY convention which has divided the 'natural' and the 'social' sciences. At the least it can be argued that the fourth IPY finally acknowledges the equal importance of studying and understanding the role of humans and society in changing environmental conditions. While the shape and degree to which the fourth IPY will impact global politics will perhaps not thoroughly be analyzed or understood for years to come, the following section – through three case studies – offers a preliminary glimpse into how global politics once again was conditioned by and conditions the parameters of the fourth IPY.

Creating legal norms: The politics of climate change science

> If we don't pay attention to the poles now, we have avoided our responsibilities as 'as explorers of the planet'. (David J. Carlson, oceanographer and Director of the IPY International Program Office based at the British Antarctic Survey in ICSU 19 October 2005)

Over the past half-century biologists and geophysicists have increasingly come to recognize the relationship between humans and environmental degradation and change (though scientific studies of this relationship date back to at least the early 1900s; see Nilsson in this volume). The consequences of these scientific findings buttressed by environmental realities on the ground have come to bear directly on global policy regarding international development. While new non-state actors have significantly shifted the rhetoric of global politics, equally so, they are changing the parameters by which international law not only operates but how it is defined altogether. Under this changing political context, international law is 'disaggregating into multiple, sometimes overlapping, lawmaking communities' where neither states nor official political leaders are at the centre (Levit 2006, 4).

International law as such, rather than dictated by state compliance – or sovereignty as has been the tradition of international law – now includes a wide range of new stakeholders who are creating, interpreting and enforcing new informal rules and normative behaviours, which often eventually become subsumed and assumed under formal rules of law (Levit 2006, 1). These non-traditional, informal rules are what Levit views as 'bottom-up transnational lawmaking' and are driven by ideological concerns of those dealing with 'the day-to-day technicalities of their trade' (Levit 2006). Climate change is reminiscent of Levit's bottom-up transnational lawmaking as a host of non-state local, regional, and transnational efforts are being forged to try and mitigate further human impacts on the natural environment.

The history of the IPYs also reflects these shifting legal attitudes regarding the human/nature relationship and the inclusion of new actors in law-making processes. Most particularly, IGY science during the 1950s, while expanding the Cold War to outer space, was ironically also responsible for fostering the first international treaty designating an entire continent as common heritage to be used solely in advancing the quest for scientific knowledge. The 1959 Antarctic Treaty, which entered into force in June 1961, mandated Antarctica to remain nuclear free, its environment protected, and as a means for furthering international scientific cooperation. The Antarctic Treaty's adoption (For more see Jabour and Haward; Rothwell in this volume) was a significant shift from traditional state sovereignty in that Antarctica became a 'special conservation area' with no one particular state maintaining sovereignty over the continent. Shortly following the Antarctic Treaty, in 1967 Malta's United Nations Ambassador, Arvid Pardo, proposed to the General Assembly that the resources of the sea should be declared the 'Common Heritage of Mankind'. This instigated negotiations resulting in the 1982 UN Convention on the Law of the Sea (UNCLOS) which, in effect, strengthened a growing international normative belief that protecting the world's environment begins with international rights and obligations rather than domestic state sovereignty.

The 1987 Brundtland Report helped further the belief in global interdependence by setting a formal international pledge to understand and address the human impacts on the world's natural environment. Five years later at the UN Conference on Environment (Rio Summit) the United Nations Framework Convention on Climate Change was adopted. While the idea of 'common heritage' had lost its prominence by the time of Rio (replaced by the preferable discourse of common concern for mankind) (Bernstein 2000: 472), beyond bringing climate change into the formal agenda, Rio also expanded the context in which global politics had by that time begun to operate. The Rio Declaration called for 'a new and equitable global partnership through the creation of new levels of cooperation among states, key sectors of societies and people ... to conserve, protect, and restore the heath and integrity of the Earth's ecosystem ... [and establish] environmental measures [to] address transboundary or global environmental problems' (UNEP 1993, vik. 1,3, 4 5 in Held 2003, 171). As such, Rio sedimented a new norm of non-state political participation.

Since Rio, global climate change has come to consume an ever-increasing proportion of the discourse of sustainable development. The vast consensus regarding the human impact on climate change has been a process constitutive of its accompanying discourse; a discourse which has become contested knowledge for an increasing number of political players including politicians, scientists, and a wide variety of other non-state political actors. Which actors have the right to determine this scientific knowledge, how this knowledge is used – much less what scientific issues are considered of greatest importance – are continuously contested processes which are played out in the intersection between politics and science. Political institutions such as the European Union (EU) and the World Wildlife Foundation, for example, have become pivotal intermediaries for translating the discourse of scientific research into particular policy recommendations and prescriptions. And it is through this process that the construction of knowledge becomes a sedimented reality. According to historian of science Jan Golinksi:

> As facts are translated from the language in which they are represented among specialists to language appropriate for a lay audience, they become consolidated as knowledge. As experts describe their findings to nonexperts, facts are simplified and rendered more dramatic, and the sureness with which they are held is strengthened, even among the experts themselves ... 'certainty simplicity, vividness originate in popular knowledge ... Therein lies the general epistemological significance of popular science' (Fleck, Ludwik in Golinski 2005, 34).

While this relationship between science and politics is not a new phenomenon or even a newly discovered reality, the sphere in which climate change debates are taking place, whose scientific knowledge is considered legitimate, and how and in what shape the results are turned into political action mark a significant shift in classical assumptions regarding state sovereignty and resource development and the way in which international law traditionally proceeds. Increasingly so, the scientific community has grown to play a decisive role in defining the conditions upon which climate change is discussed and understood. Policy prescriptions regarding sustainable development are significantly dependent on the technical expertise, research, and assessments of the scientific community. As such, prompted by a global intersubjective consensus of the human desire to sustain life on earth, it could be questioned as to whether or not there is an inherent call for political action embedded in climate change science and in and of itself and if science, as such, is leading the global policy debate, directly helping to steer and guide the international legal changes taking place alongside these discussions. Have scientists always served as creators and enforcers of international politics or we witnessing the beginning of something distinctly unique since the onset of Westphalia? One robust example of this stems from a comment by Susan Salomon, senior scientist at the US National and Oceanic Atmospheric Association (NOAA). After serving as a contributing researcher and author of the Fourth

Assessment Report of the Intergovernmental Panel on Climate Change (IPCC), when questioned by the media as to what this should mean for policy makers, she declared that as a scientist her job is not to engage in political debate: 'I can only give you something that's going to disappoint you, sir, and that is that it's my personal scientific approach to say it's not my role to try to communicate what should be done ... I believe that is a societal choice. I believe science is one input to that choice, and I also believe that science can best serve society by refraining from going beyond its expertise' (Solomon in Revkin, 6 February 2007).

Is the Intergovernmental Report on Global Climate Change merely a scientific document or is its research agenda political by its very existence and mandate? More generally, is science ever abstracted from the social world? This ongoing, yet shifting relationship between law, politics, and science beginning with Brundltand and reinforced at Rio is reflected in the fourth IPY. The fourth IPY not only focuses on the need to understand the natural world, but further takes into account the holistic (and interdisciplinary) turn in the sciences for understanding the human impacts of past eras of industrial development. Likewise, it focuses on emerging trends towards sustainable management approaches that highlight the interdependence between state, individual and collective rights and responsibilities.

As climate change science brings increased attention to the polar regions (the poles as the world's barometer), the mandate of the IPY as stated by the International Council for Science (ICSU) has constructed into Arctic science policy the global scientific convictions of the relationship between human development and the environment as reflected in its mission: 'at a time when humans are exerting an increasing impact on the planet, and when the human condition is increasingly affected by global changes, the polar regions are especially important and relevant' (ICSU IPY 2007–2008 Planning Group November 2004, 9). Likewise, the IPY seeks to 'accelerate progress towards providing the required policy-relevant answers' (ICSU IPY 2007–2008 Planning Group November 2004), answers which assume that in order to achieve sustainable development practice (much less how it is defined) necessitates bi-lateral, multilateral, yet moreover, global coordination including a wide range of political actors – which combined are shifting the parameters by which international law operates.

While it is anticipated that the research findings and prescriptions of the fourth IPY will bring about a better understanding of climate change, perhaps inextricably attached to these findings and the increased public awareness accompanying them will be a heightened sentiment to take political action. In this case, something can be said of the political agency of the scientific community in affecting the course of global politics.

Indigenous knowledge as governance: Remaking the boundaries of science

We, the Inuit are true examples of what Charles Darwin came close to, namely "the survival of the fittest." Throughout history we have developed technologies, techniques, and know how adapted to the needs that makes us survive as the fittest within the Arctic environment…Our aim is to preserve the best of the old and adopt the best of the new (Jakobsen 15 April 1997, 3).

Legally speaking, the ideological underpinning of Westphalian international law is liberal institutionalism. The boundaries of this liberal legal framework have shifted over time, rearticulating who is included and precluded as legitimate international actors. During colonization, indigenous peoples –representing the symbolic 'other' – helped construct the boundary between legitimate political actors and the world outside the formal parameters of the system. In recent years, however, alongside a broader increase in non-state political agency, indigenous peoples in particular, in certain capacities, have come to constitute a formal attribute of the 'inside' – a process which has reconceptualized the political space of international politics (Shadian 2006). The political agency of indigenous actors has more generally become part of a larger transformation of international law in which indigenous rights have become a definitive piece of two interdependent strands – human rights and international development.

During the colonial era, indigenous conceptions of stewardship approaches toward governing land ownership and resources functioned as the basic justification for European expansion, undermining any existing indigenous self-determination. Since this time, the long trumped indigenous principle of stewardship has resurfaced within international policy discourse. Rather than serving as the means by which political leaders ignore or override indigenous autonomy, stewardship has become central to sustainable development discourse, and within this, the means by which indigenous leaders have justified their claims for self-determination. This has been accomplished through the 'right' to participate in political processes rather than through ownership over the land as sovereign states (Shadian 2006). According to one Inuit leader the Inuit concept of stewardship means that:

> Inuit have always lived on that land and used it. Inuit have a right to the land because of their heritage. This is the foundation for the legal concept, or meaning of 'aboriginal rights'. … Aboriginal rights, in theory, are property rights, that is, the recognition of ownership of land and the people who have lived on and used that land from the beginning of time. Native people are to be guaranteed the right to use that land (Yabsley, October 1976, 22, 50).

The policy area in which indigenous stewardship has found its re-invention is within the politics of sustainable development and central to this move is the recognition of traditional indigenous knowledge as a critical body of knowledge in ensuring human survival. More recently, the discourse of sustainable development has further shifted to include resilience and adaptation and indigenous groups in the Arctic similarly argue that resilience and adaptation is an ongoing reality that Arctic indigenous peoples have lived with throughout their cultural history.

In general terms, traditional indigenous science as a concept is predicated on a holistic approach to science that links the natural and social world. Indigenous knowledge can be defined as a system of knowledge 'associated with a fixed territorial space for a considerably long period of time' (Fernando 2003, 56). The underlying premise of indigenous knowledge is that it is particular to a specific community, culture, or society and serves as the basis by which local communities are constructed and operate. Such forms of community structures include agriculture, natural resource management (including fishing and hunting), community well-being (health care), spiritual beliefs, and other cultural practices (Fernando 2003).

In the Arctic, indigenous groups view traditional indigenous science as the knowledge which has provided indigenous peoples with the ability to survive and sustainably manage the Arctic's land and resources since time immemorial (AHDR 2004). In Arctic policy, an intersubjective consensus has evolved into that where indigenous knowledge is taken to be a key component in the actual definitions regarding sustainable development and resilience and adaptation (see Arctic Council Framework Document for the Sustainable Development Programme and Intergovernmental Panel on Climate Change 2008). In terms of operationalizing indigenous knowledge (often viewed as the binary 'other' to 'Western' science), the aim is to turn local tradition into a method for doing 'scientific' practice – or otherwise doing sustainable development. Most often this is facilitated through projects and policies calling for the creation of databases for preserving indigenous tradition. The IPY science plan for 2007–08 is one example. According to the ICSU/WMO Joint Committee for IPY 2007–08:

> 2007–2008 will become a true milestone in polar studies … engagement of polar residents, including polar indigenous people, in research planning, observation, processing and interpretation of the various data sets created through IPY projects … and as a vital component of the data collection, monitoring, data analysis and data management processes … also, increasingly, to many projects undertaken by scientists from physical and biological disciplines … [and] … at least as important, are the projects that are initiated and conducted by polar communities and regional organizations, involving their own knowledge and observations of local processes and phenomena (Allison et al. February 2007, 51).

Arun Argwal, however, argues that efforts such as the IPY to classify, document, and preserve indigenous knowledge is, in and of itself, operating on the assumption that knowledge is static – when all knowledge systems are in fact dynamic. As such, through the processes of making indigenous knowledge databases, the very concept of indigenous knowledge itself becomes a body of knowledge which will not be traditional but rather new as it is created according to the epistemological assumptions of Western scientific method. According to Argwal, to begin with the idea that indigenous knowledge is local and then an attempt to define a certain cultural means of knowledge production through an alternative, culturally produced lens (which assumes science to be universal), is ontologically problematic. Jude Fernando reifies this sentiment:

> ... the current interest in [Indigenous Knowledge] IK is largely a result of a force external to where they are located. Therefore ... it is far more productive to view IK as a social phenomenon produced within a specific social, economic, and political context and thereafter, proceed to analyze the relevance of such meanings and the institutional and power relations embedded in them ... (Fernando 2003, 58).

Alongside these arguments, however, there is another body of literature which asserts that the transfer of many forms of knowledge across cultures throughout history is nothing new and often times creates new sets of hybrid knowledge (Fernando 2003, 58). This most recent turn towards a focus on indigenous knowledge – rather than being considered a cultural artefact in the form of a database – can perhaps be better perceived as another example of hybrid forms of knowledge. Moreover, the analytical interest of indigenous knowledge – while perhaps a constructed idea at the outset – is the political forces intertwined within the processes including who defines this knowledge, the power which it affords and likewise who is able to speak on its behalf.

Fernando, for example, analyzes a case study of one particular development NGO, which carried out a project in an indigenous community in Sri Lanka based on the idea of promoting indigenous knowledge. In this case, Fernando explains how the idea of indigenous knowledge became appropriated first, by the NGO as a means to attract investment and then later by multinational companies through the commodification of this knowledge. In the Arctic, however, a very different narrative has emerged. As indigenous resource development is taken to be something practised since time immemorial, indigenous knowledge has become considered the key to understanding and creating sustainable development practice. As such, in the Arctic, indigenous knowledge has become a focus of sustainable development policy as well as a significant means by which indigenous peoples have legitimized a stake, claims to and control over local research and development practices. Through the evolution of Arctic policy (within the Arctic Council as well as nationally and locally), the role and meaning of Arctic science has evolved as an amalgamation of the rise of indigenous political agency acting in concert with broader global sustainable development policy and practice. This interaction between indigenous agency and sustainable development policy has helped shape how the Arctic as a region is defined, the scientific basis upon which knowledge is constructed and who is afforded political legitimacy.

At the international level – whether only in rhetoric or actual practice – indigenous knowledge has translated into a 'real' science through certain initiatives. Such policies include the International Labour Organization (ILO), the Food and Agriculture Organization (FAO), the World Bank, and ICSU (International Council of Science). In the Arctic in particular, this includes the Arctic Council and, specific to the interests of this chapter, it is also embedded in the aims and mission of the fourth IPY. This move to include traditional indigenous knowledge is restructuring the ways in which science in the polar regions is defined and practised. However, indigenous science likewise, remains bound by the ability to construct an indigenous discourse which speaks to existing norms of 'traditional' Western scientific definitions and practice which R. Andolina et. al. (2005) refer to as 'developmentally appropriate culture' – a combination of self-made 'ethno' policies and neoliberal institutions including the World Bank and state international development agencies such as the United States Aid for International Development (USAID) and the Canadian International Development Agency (CIDA). As such, the very language upon which traditional indigenous knowledge has gained its legitimacy is by finding a space within the existing scientific frameworks while simultaneously shifting the boundaries defining legitimate science. Referring again to the IPY science plan:

> projects that are initiated and conducted by polar communities and regional organizations, involving their own knowledge and observations of local processes and phenomena. ... will greatly increase through IPY 2007–2008 to include the sustainable use of local resources, for example, in fisheries, exploitation of reindeer/caribou populations and environmental-friendly tourism; indigenous cultural and language sustainability; increased resilience of local economic and social systems through co-management, local self-governance and information exchange among local stakeholders; and interactions with the ongoing industrial development of the polar regions, including monitoring of local environmental and social impacts, primarily in oil and gas, and other mineral exploitation (Allison et al. February 2007, 51).

A case in point of this co-dependent process in remaking the boundaries of Arctic science and constructing a 'developmentally appropriate culture' is the Nunavut Research Institute in Canada. In 1993, the Nunavut Act was approved and put into effect in 1999. The Act was the resolution of a Canadian Inuit land claim settlement which ended 20 years of negotiations (breaking off Nunavut from the Northwest Territories). The Nunavut Research Institute was created in 1995 and over the course of time has shifted the parameters for conducting science in a vast area of Canada. The Institute's mission is 'to provide leadership in developing, facilitating, and promoting traditional knowledge, science, research and technology as a resource for the well-being of people in Nunavut' (Nunavut Research Institute home page). In 1988, the Northwest Territories Scientists Act was passed requiring all researchers interested in conducting research in the region to acquire a license or permit. Through the Nunavut Act, the decision and administration of permits has shifted to the Nunavut Research Institute. Of the three types of applications available, two of which being health and land/water issues, the third regards social sciences and traditional knowledge research. This system of licensing, while falling neatly within the parameters of dominant Canadian institutional culture, has also provided Inuit increasing power to help define and set new boundaries for carrying out Arctic science in Nunavut.

While the Arctic's indigenous communities may have traditionally played an important role in IPY research, in formal terms they were considered part of the Arctic landscape and likewise the objects of science rather than as scientists possessing knowledge of the Arctic ecosystem. The recent agency of indigenous political actors has turned the traditional knowledge base of the Arctic's indigenous peoples on its head. Rather than considered to be lacking method, scientific rigor or results, traditional indigenous knowledge has created a new politics for practising science. Whether or not local indigenous traditions will be compromised in the name of documentation and the need for data, the role of indigenous science in the fourth IPY is something new and in a growing number of instances unavoidable for those who carry out Arctic research (as this instance illustrates). These changes will undeniably leave a lasting legacy on the practices of Arctic science in general and the IPY in particular.

Big Science from government to governance: From private sector cooperation to coordination

> In order to strengthen international science for the benefit of society, ICSU mobilizes the knowledge and resources of the international science community to … [p]rovide independent, authoritative advice to stimulate constructive dialogue between the scientific community and governments, civil society, and the private sector (ICSU IPY 2007–2008 Planning Group 2004).

On 2 August 2007, Russia staged a global media event when its most famous Arctic explorer planted a flag on the outer continental shelf of the North Pole – a shelf which the government argues extends outward from Russia. The reaction which ensued was a reawakening to the strategic importance the Arctic for the eight Arctic countries (e.g. Shadian 2009; Borgerson 2008; Huebert 2008 and see Huebert in this volume). Merely in terms of science, immediate responses included joint research expeditions from the Arctic countries including Sweden, Denmark, and the United States. Likewise, the flag ceremony initiated an international debate comprised of environmental groups, fisheries, shipping industries, scientists, political pundits, and the media over who has the authority to stake a claim in the future course of Arctic development and under what conditions. The Russian flag planting itself, it could be argued, did as much to advance both public and scientific interest in the polar regions as did the kickoff for the fourth IPY.

Yet, by the time Russia made front page news in August 2007, growing interest over the potential impacts that increased global climate change may have for Arctic resource development was long underway. In 2000, the United States Geological Survey (USGS) conducted an Arctic assessment of the potential for undiscovered and – in the event of an ice-free sea – technically recoverable oil and gas resources.

Following collaboration with the Geological Survey of Denmark and Greenland (GEUS), the USGS in 2007 completed another survey of potential Arctic oil and gas reserves in the East Greenland Rift Basins Province. The assessment took place in Northeast Greenland because it is considered an archetype location for a Circum-Arctic Resource Appraisal (CARA). According to the USGS, the area has similar important characteristics with many Arctic basins including 'sparse data, significant resource potential, great geological uncertainty, and significant technical barriers to exploration and development' (Gautier 2007, 4). The 2007 USGS Circum-Arctic Resource Appraisal was the 'the first systematic and comprehensive analysis of the undiscovered petroleum resources of the Circum-Arctic in the public domain' (Robertson 28 August 2007). According to USGS Director Mark Myers, uncovering the potential for resource exploitation in the Arctic – an area which he points out as also being environmentally sensitive, maintaining technological risk and geological uncertainty will be 'critical to our understanding of future energy supplies to the United States and the world' (Robertson 28 August 2007). These USGS survey findings have since become one of the most often cited preludes to any discussion regarding the future course of Arctic development.

While the fourth IPY is most clearly interested in better understanding the impacts of global change on the natural environment and human societies, a changing Arctic climate is also impacting the global geopolitics of resource development. The potential impacts that climate change may impose for future energy security demands are less discussed in the fourth IPY yet are nonetheless of great interest. The US Geological Survey itself has an official IPY project entitled 'Petroleum Resource Assessment of Arctic'. The project, which includes cooperation with British Petroleum and StatoilHydro is a public/private project which among other aims includes an 'assessment of energy resources in the circum-arctic area including, oil, gas, coalbed methane and methane hydrates' (IPY Activity ID No: 86 Updated on 28 March 2008). According to Suzanne Weedman of the USGS:

> This is very much a part of what we do. Our responsibility is to assess the undiscovered oil and gas using geological information ... If you look at the objectives of International Polar Year, one of them is to assess the impact of these changes on people who live in the Arctic. Knowing about the energy resources might be very interesting because there is the potential of development in the Arctic. That's not for us to decide, but it is the reality (Adam 2008).

While the fourth IPY is said to be unique in many ways, including the introduction of the human dimension and social sciences, it also marks a distinct break from IPY tradition through the formal inclusion of private industry in IPY research proposals. Another example of private funding is the IPY project entitled 'Impact Assessment with Indigenous Perspectives' (IPY Activity ID No: 378 Updated on 28 March 2008). The project is a course aimed to 'prepare Indigenous students for environmental policy-and decision-making and management, and to provide them with a deeper understanding of the linkages between economic, social and ecological systems'. It is a joint public/private project which includes funding from the Sámi University College, StatoilHydro, University of the Arctic, and the Norwegian Ministry of Foreign Affairs (IPY Activity ID No: 378).

The point here is not to debate the merits of private funding of science. Rather, formal private investment, such as the two IPY cases here, is highly non-traditional in Big Science. During the IGY private industry was very much a part of IGY research – the military industrial complex was in large part the merging of public and private science as the public often paid for private military research. Yet, whereas the IGY was public funding for military science in which the products were assumed to provide governments the means to protect the 'national' security of the people, the inverse is the case in the fourth IPY. In the two projects mentioned here private industry is funding public science in order to produce knowledge directly for the benefit and use of the private sector (though it can be argued that the Norwegian government has a sizeable investment in StatoilHydro). What then are the implications of these changes? What is at stake? Is bringing in more participants into research the democratization of science or is it bringing science out of the 'official' public into the 'private' sphere? Is private funding of science undermining public input into science or are the spaces of science shifting and therefore redefining the definitions traditionally contained within the conceptual boundaries of public and private science? The role of private industry in the fourth IPY undoubtedly offers an initial glimpse into these very interesting questions.

Conclusion: IPY science for whom and by whom?

Generally speaking, new political eras throughout history often usher in new belief systems which, while new in many aspects, always contain legacies of past practices and the structures of enduring intellectual thought. The last IPY – IGY took place at the height of a world dominated by state power tensions in which all other politics were held at bay and domestic state ambitions were carried out in the name of science. Instead, the fourth IPY could be said to fall in the midst of making a new global political framework. As the world's attention remains captured by uncertainties of climate change as well as future global energy supplies, the fourth IPY both highlights the lingering legacies of Westphalia – including the ongoing tradition of state funded projects and collaboration. Likewise, the latest IPY also exemplifies the interface between the ongoing legacy privileging the state in Big Science (state science funding) intertwined with the influences of new actors carrying out IPY projects and controlling the flow of information. From a historical perspective, the fourth IPY exemplifies the increased power and formal role of not only private industry but a broad range of new non-state actors to define what constitutes appropriate science and the ways in which science is carried out. By additionally focusing on indigenous communities as scientific experts this chapter is also one case in point of increasing instances in which the state serves as only one of many actors affecting the creation and control over the flow of intellectual knowledge. Likewise, this chapter questions the larger role of science in steering and guiding international legal change bringing to question whether the scientific community is becoming a political entrepreneur in its own right.

In terms of global political and legal trends and changes, the implications of the fourth IPY – the largest international scientific undertaking to date, which included thousands of scientists, over 200 projects and the involvement of more than 60 nations – remains in question. Adorned with scientists who are considered legitimate authority figures with a mandate to help better understand and assess global climate change in a political world where resource development and economic growth remain fundamental to global political processes, the scientific evidence produced from the fourth IPY will most certainly contribute new local, regional, and global policies.

At its very least, the fourth IPY will be remembered for the moment in which indigenous peoples were formally written into the histories of the IPY and its legacies. Likewise, the fourth IPY is also unique in that, while individual science projects are aimed at understanding the human impact on the earth, these same projects require not only the backing of governments (states are the overriding institutions financing science) but also include private investment (whose interests are never endogenous to the larger narrative). Lastly, the future discovery, exploitation and ownership of the Arctic's resources accompanied by an emerging reality of new routes for shipping them to market is creating a new global debate over the future of the Arctic policy, security and development. As the future decisions to be made are as much about science as they are about politics, the results of the IPY have a timely potential to contribute as much to the future of Arctic policy as do political leaders themselves. If viewed in this light, the fourth IPY serves as a potent tell-tale sign of the dominating paradox of governance in Haas' non-polar world and the emergence of new relationships between the role of science and global politics. Given these stakes, perhaps the lasting historical narrative of the fourth IPY will be written as the IPY which set out to understand global climate change and, occurring at a critical juncture in the history of Arctic and global politics, it was also a noteworthy player in the history of a changing Arctic politics and the overall narrative of an emerging post-Westphalian system.

References

Abele, F., Courchene, T., Seidle, L. and St-Hilaire, F. (eds) (forthcoming), *Northern Exposure: Peoples, Powers and Prospects for Canada's North* (Canada IRPP Art of the State series) <http://www.irpp.org/indexe.htm>, accessed 8 October 2008.

Adam D. (no date), 'Scramble is on for Arctic oil', *Guardian Weekly* [Online] <http://www.guardian.co.uk/guardianweekly/story/0,,1756812,00.html>, accessed 18 July 2008.

Agnew, J. (2005), 'Sovereignty Regimes: Territoriality and State Authority in Contemporary World Politics', *Annals of the Association of American Geographers* 95:2, 437–61.

AHDR (ed.) (2004), *Arctic Human Development Report* (Akureyri: Stefansson Arctic Institute).

Allison, I., Béland, M., Alverson, K., Bell, R., Carlson, D., Danell, K., Ellis-Evans, C., Fahrbach, E., Fanta, E., Fujii, Y., et al. (2007), 'The Scope of Science for the International Polar Year 2007–2008', *ICSU/WMO Joint Committee for IPY 2007–2008* [Online] <http://216.70.123.96/images/uploads/LR*PolarBrochur eScientific_IN.pdf>, accessed 18 July 2008.

Anderson, B. (1991), *Imagined Communities: Reflections on the Origin and Spread of Nationalism* (London: Verso).

Andolina R., Radcliffe S., and Laurie, N. (2005), 'Development and Culture: Transnational Identity Making in Bolivia', *Political Geography* 24, 678–702.

Ansell, C. and Weber S. (January 1999), 'Organizing International Politics: Sovereignty and Open Systems', *International Political Science Review* 20:1, 73–93.

Archer, C. and Scrivener, D. (2000), 'International Co-operation in the Arctic Environment', in Nuttall and Callaghan (eds).

Archibugi, D. et al. (eds) (1998), *Re-Imagining Political Community: Studies in Cosmopolitan Democracy* (Cambridge: Polity Press).

Arctic Council (2000), 'Arctic Council Framework Document for the Sustainable Development Programme' [Online] <http://arctic-council.org/filearchive/Framework%20Document.pdf>, accessed 7 October 2008.

Beach, H. (2001), *A Year in Lapland: Guest of the Reindeer Herders* (Washington: University of Washington Press).

Bernstein, S. (December 2000), 'Ideas, Social Structure and the Compromise of Liberal Environmentalism', *European Journal of International Relations* 6, 464–512.

Betsill, M. and Bulkeley, H. (2004), 'Transnational Networks and Global Environmental Governance: The Cities for Climate Protection Program', *International Studies Quarterly* 48, 471–93.

Borgerson, S. (2008), 'Arctic Meltdown', *Foreign Affairs* 97:2 [Accessed 2 October 2008, from ABI/INFORM Global database].

Brody, H. (1990), *Living Arctic: Hunters of the Canadian North* (Washington: University of Washington Press).

—— (2002), *The Other Side of Eden: Hunter-gatherers, Farmers and the Shaping of the World* (New York: North Point Press).

Bulkeley, H. (2005), 'Reconfiguring Environmental Governance: Towards a Politics of Scales and Networks', *Political Geography* 24, 875–902.

Carlson, D. (2005), 'International Council for Science launches International Polar Year 2007–2008, an Endeavor of Historical Proportions in ICSU', Press release (19 October 2005) [Online] <http://www.scar.org/media/pressreleases/IPYreleasefinaltem.PDF>, accessed 16 July 2008.

Chapin III, F., Hoel, M., Carpenter, S., Lubchenco, J., Walker, B., Callaghan, T., Folke, C., Levin, S., Mäler, K., Nilsson, C., Barrett, S., Berkes, F., Crépin, A., Danell, K., Rosswall, T., Starrett, D., Xepapadeas, A., and Zimov, S. (2006), 'Building Resilience and Adaptation to Manage Arctic Change', *AMBIO: A Journal of the Human Environment* 35:4, 198–202.

Chapman, S. (1954), 'The International Geophysical Year and Some American Aspects of it', *Proceedings of the National Academy of Sciences of the United States of America* 40:10, 924–26.

Editor (2002), *Sigma Xi International Newsletter* 1:4 [Online] <http://www.sigmaxi.org/programs/international/newsletter1102.pdf>, accessed 12 July 2008.

Elzinga, A., Nordin, T., Turner, D., and Wrakberg, U. (eds) (2004), *Antarctic Challenges: Historical and Current Perspectives on Otto Nordenskjöld's Antarctic Expedition 1901–1903* (Gothenburg, Sweden: Royal Society of Arts and Sciences).

Fernando, J. (2003), 'NGOs and Production of Indigenous Knowledge under the Condition of Postmodernity', *The ANNALS of the American Academy of Political and Social Science* 590, 54–72.

Fleming, J. (ed.) (2004), *Proceedings of the International Commission on History of Meteorology 1.1.* (11–12 July 2001) (Mexico City: International Perspectives on the History of Meteorology: Science and Cultural Diversity).

Freeman, M.M.R. (1992), 'The Nature and Utility of Traditional Ecological Knowledge', *Canadian Arctic Resources Committee* 20(1) [Online] <http://www.carc.org/pubs/v20no1/utility.htm>, accessed 13 March 2009.

Gautier, D. (2007), 'Assessment of Undiscovered Oil and Gas Resources of the East Greenland Rift Basins Province: U.S. Geological Survey Fact Sheet 2007–3077' [Online] <http://pubs.usgs.gov/fs/2007/3077/>, accessed 18 July 2008.

Golinski, J. (2005), *Making Natural Knowledge: Constructivism and the History of Science* (Illinois: University of Chicago Press).

Haas, R. (2008), 'The Age of Nonpolarity: What Will Follow U.S. Dominance', *Foreign Affairs* 87:3, 44–56. [Accessed 2 October 2008, from ABI/INFORM Global database].

Held, D. (2003), 'The Changing Structure of International Law', in Held and A Bhabha, (ed.).

—— and A Bhabha, H. (ed.) (2003), *Nation and Narration* (London: Routledge).

Held, D. and McGrew, A. (eds) (2003), *The Global Transformations Reader: An Introduction to the Globalization Debate* (Cambridge: Polity Press).

Howarth, D., Aletta, J.N. and Stavrakakis, Y. (eds) (2000), *Discourse Theory and Political Analysis: Identities, Hegemonies and Social Change* (Manchester: Manchester University Press).

Huebert, R. (2008), 'Canada and the Changing International Arctic at the Crossroads of Cooperation and Conflict', in Abele, F. et al. (eds).

ICSU IPY 2007–2008 Planning Group (November 2004), 'A Framework for the International Polar Year 2007–2008', International Council for Science [Online] <http://www.icsu.org/Gestion/img/ICSU_DOC_DOWNLOAD/562_DD_FILE_IPY_web_version.pdf>, accessed 18 July 2008.

Intergovernmental Panel on Climate Change (2008), 'Climate Change 2007 – Impacts, Adaptation and Vulnerability: Working Group II Contribution to the Fourth Assessment Report of the IPCC' (Cambridge: Cambridge University Press).

IPY (2008a), 'Full Proposals for IPY 2007–2008 Activities: (Activity ID No: 86)' [Online] <http://classic.ipy.org/development/eoi/proposal-details.php?id=86m>, accessed 18 July 2008.

IPY (2008b), 'Impact Assessment with Indigenous Perspectives: (Activity ID No: 378)' [Online] <http://classic.ipy.org/development/eoi/proposal-details.php?id=378>, accessed 18 July 2008.

Jackson, P. (2008), 'Pierre Bourdieu, the "Cultural Turn" and the Practice of International History', *Review of International Studies* 34, 155–81.

Jackson, R.H. and James A. (eds) (1993), *States in a Changing World: A Contemporary Analysis* (Oxford: Clarendon).

Jacobsson, M. (2004), 'Acquisition of Territory at the Time of Otto Nordenskjöld: A Swedish Perspective', in Elzinga, A. et al. (eds).

Jakobsen, A. (1997), Presentation: 'One Day Symposium on Wildlife Management, Indigenous Peoples and the World of Politics' (Brussels).

Keskitalo, C. (2002), 'Region-building in the Arctic: Inefficient institutionalism?: A Critical Perspective on International Region-building in the "Arctic"' (A paper prepared for presentation at the ISA annual convention. New Orleans, 25–27 March 2002) [Online] <http://www.isanet.org/noarchive/keskitalo.html>.

Krasner, S. (ed.) (1983), *International Regimes* (Ithaca: Cornell University Press).

—— (1993), 'Economic Interdependence and Independent Statehood', in Jackson and James (eds).

—— (1995–1996), 'Compromising Westphalia', *International Security* 20:3, 115–51.

—— (1999), *Sovereignty: Organized Hypocrisy* (Princeton: Princeton University Press).

—— (2001), *Problematic Sovereignty* (New York: Columbia University Press).

Krupnik, I., Bravo, M., Csonka, Y., Hovelsrud-Broda, G., Müller-Wille, L., Poppel, B., Schweitzer,P., Sörlin, S. (2005), 'Social Sciences and Humanities in the International Polar Year 2007–2008: An Integrating Mission', *Arctic* 58: 1, 91–101.

Kuptana, R. (1996), 'Statement to the Second Conference of the Parties to the Unites Nations Framework on Climate Change: Inuit Perspectives on Climate Change' (Geneva).

Latour, B. (1999), *Pandora's Hope: Essays on the Reality of Science Studies* (Cambridge: Harvard University Press).

Levit, J. (2006), 'International Law Happens: Executive Power, American Exceptionalism, and Bottom-up Lawmaking', *bepress Legal Series* Paper 1676. [Online] <http://law.bepress.com/expresso/eps/1676>, accessed 18 July 2008.

Luedecke, C. (2004), '6. The First International Polar Year (1882–1883): A Big Science Experiment with Small Science Equipment', Fleming (ed.).

Martin, D. (1958), 'The International Geophysical Year', *The Geographical Journal* 124:1.

NASA (2008), 'NASA's Origins and the Dawn of the Space Age: Monographs in Aerospace', 10 [Online] <http://www.hq.nasa.gov/office/pao/History/monograph10/korspace.html>, accessed 16 March 2008.

NOAA Arctic Research (2008), 'Office: The First International Polar Year: The Arctic Environment in Historical Perspective', [Online] <http://www.arctic.noaa.gov/aro/ipy-1/>, accessed 20 July 2008.

NOAA Staff (2007), 'NOAA Celebrates 200 Years of Science, Service and Stewardship', *Feature Stories: The International Geophysical Year* [Online] <http://celebrating200years.noaa.gov/about.html>, accessed 16 June 2008.

Nowotny, H., Scott, P., and Gibbons, M. (2003), 'Introduction: Mode 2 Revisited: The New Production of Knowledge' *Minerva* 41, 179–94.

Nunavut Research Institute home page, <http://www.nri.nu.ca/iqaluit.html>, accessed 7 October 2008.

Nuttall, M. and Callaghan, T. (eds) (2000), *The Arctic: Environment, People, Policy* (Australia: Hartwood Academic Publishers).

Osherenko, G. and Young, O. (2005), *The Age of the Arctic: Hot Conflicts and Cold Realities* (Cambridge: Cambridge University Press).

Osiander, A. (2001), 'Sovereignty, International Relations, and the Westphalian Myth', *International organization* 55:2, 251–87.

Pestre, D. (2003), 'Regimes of Knowledge Production in Society: Towards a More Political and Social Reading', *Minerva* (Netherlands: Kluwer Academic Publishers).

PotoÄnik, J. (2007), 'Polar Environment and Climate: The Challenges', *European Research in the Context of the International Polar Year European Commissioner for Science and Research* [Online] <http://www.europa-nu.nl/9353000/1/j9vvh6nf08temv0/vhisn5yeiuzx?ctx=vgo02ttv3ezn&start_tab0=20>, accessed 15 July 2008.

Revkin A. (2007), 'Scientist at Work: Susan Solomon Melding Science and Diplomacy to Run a Global Climate Review', *New York Times* [Online]

<http://www.nytimes.com/2007/02/06/science/earth/06profile.html>, accessed 16 July 2008.

Roberts, G. (2006), 'History, Theory and the Narrative Turn in IR', *Review of International Studies*, 32, 703–14.

Robertson, J. (2007), 'News Release: USGS Releases New Oil and Gas Assessment of Northeastern Greenland', *USGS*. [Online] <http://www.usgs.gov/newsroom/article_pf.asp?ID=1750>, accessed 18 July 2008.

Rudolph, C. (2005), 'Sovereignty and Territorial Borders in a Global Age', *International Studies Review* 7:1, 1–20.

Ruggie, J. (1993), 'Territoriality and Beyond: Problematizing Modernity in International Relations', *International Organization* 47:1, 139–74.

Sagdeev, R. and Logsdon, J. (2008), *United States–Soviet Space Cooperation during the Cold War* [Online] <http://www.nasa.gov/50th/50th_magazine/coldWarCoOp.html>, accessed 16 July 2008.

Schwartz, C. (1996), 'Political Structuring of the Institutions of Science', in Laura Nader (ed.) *Naked Science: Anthropological Inquiry into Boundaries, Power And Knowledge* (United States: Routledge).

Shadian, J. (2006), 'Reconceptualizing Sovereignty through Indigenous Autonomy: A Case Study of Arctic Governance and the Inuit Circumpolar Conference', PhD Thesis: University of Delaware.

—— (April 2009), 'Building Bridges (and boats) Where There Was Once Ice: Adopting a Circumpolar Approach in the Artic *Policy Options*.

Skolnikoff, E. (1994), *The Elusive Transformation* (Princeton: Princeton University Press).

Sorlin, S. (2007), 'Microgeographies of Authority: Tarfala Glaciology, Saami Entrepreneurship, and the Site-bound Tradition of Climate Scepticism, ca 1940 to 1970', Paper presented at the IPY Field Stations Meeting (Scott Polar Research Institute, University of Cambridge).

Stefansson, V. (1943), *Friendly Arctic* (London: Macmillan and Company).

Yabsley, G. (1976), *Inuit Today* 5, 22–50.

Young, O. (1992), *Arctic Politics: Conflict and Cooperation in the Circumpolar North* (United States: Dartmouth).

—— (1998), *Creating Regimes: Arctic Accords and International Governance* (New York: Cornell University Press).

—— (1999), *Governance in World Affairs* (New York: Cornell University Press).

Chapter 3

Science, Cooperation and Conflict in the Arctic Region

Rob Huebert

One of the most confounding (and interesting) elements surrounding the scientific study of the polar regions is the assumption by most scientists that science is value free. Science is widely assumed to discover 'truth' and therefore has no value of its own. However, even a precursory examination of the conduct of scientific research in the polar region shows that this is not true. The debates that developed over the value of traditional knowledge and its role as a 'science' clearly demonstrated that this was not the case. But even more telling has been the reaction of the international community to the scientific discovery of different types of environmental degradation in the polar regions.

Perhaps even more important, science itself is assumed to be the means by which international cooperation can be fostered. The assumption that underlies most of the efforts to justify international scientific cooperation in the Arctic and Antarctica is based on the belief that if countries can only come together to conduct 'science', then improved international cooperation will follow. The IPY of 1957–58 is often credited with 'paving the way' to the Antarctic Treaty (IPY 2006). Likewise, this nexus between science and international cooperation forms one of the core assumptions for the International Council for Science (ICSU 2004) when it was developing the framework for the development of the IPY in 2007–08.

However, does it automatically follow that 'science', and especially the natural sciences, actually provides a universal language of cooperation? Does the pursuit of 'science' in the polar regions lead to the creation of improved international relations and cooperation? Or is it possible that this is a belief that is used to justify the call for international cooperation in the conduct of scientific research? And if indeed this is the case that international cooperation in the conduct of scientific research leads to improved international cooperation, then are there certain types of research that are more powerful in creating and fostering international cooperation? Ultimately, are events such as the IPY a powerful force in improving international relations? Or are such events important in the production of new science but relatively ineffective as tools for improving international cooperation?

In order to address these issues, the first question that needs to be asked pertains to the use of science as an organizing device. What does scientific cooperation look like, and what has it done historically to improve international cooperation? The question that this section will examine is why there is such a different set of reactions to what are ultimately scientific studies of the environmental degradation of the polar region. Why does the international community respond cooperatively in one instance while in others adopts a much more unilateral approach? Both responses are to what 'science' tells us about the impact of environmental degradation of the polar regions. Yet the responses are very different – why? What does this tell us about the nature and impact of science?

Science, CFCs and POPs in the polar regions

Prior to the mid 1980s, the polar regions were assumed to be pristine and environmentally 'pure' regions. This assumption was based on the distances these regions were from population centres and the impacts of industrialization (and the fact they 'looked' clean). However, already beginning in the 1970s scientists discovered that there was a significant and growing problem in regards to a phenomenon known as trans-boundary pollution. In the Arctic the major issue focused on the importation into the region of a host of pollutants commonly referred to as persistent organic pollutants (POPs) – mainly fertilizers and pesticides.[1] In Antarctica the main identified problem was chlorofluorocarbons (CFCs) – mainly used as a coolant and for other aerosol usages – which resulted in the reduction of the ozone level.[2] In both instances, these pollutants originated in locations far from the poles, yet were ultimately carried vast distances to be deposited at the farthest northern and southern latitudes.

As the nature of the problem became known, the immediate reaction was one of concern and then ultimately international cooperation. The solution pursued was that of international diplomacy. Ultimately two international treaties – the 1987 Montreal Protocol on Substances that Deplete the Ozone Layer and the Stockholm Convention on Persistent Organic Pollutants – were negotiated and brought into force that reduced the production and use of these chemicals.

In both instances it was the work of scientists who initiated the process by providing findings that were both highly controversial and alarming at the time of their release. Despite facing substantial opposition from interests opposed to taking action, scientists' warnings mobilized international action. Many of the first warnings were denounced as 'alarmists' or 'unsubstantiated'. However, as more and more research began to confirm the initial findings, an international consensus emerged that action needed to be taken. This action then was then translated into international treaties. These treaties have since reduced the environmental problems that were caused by the POPs and CFCs. In both instances there were substantial costs associated with the elimination of the chemicals. However, the

---

1   The breakthrough research on this was completed by Eric Dewailly and colleagues through Laval University (see Dewailly and Furgal 2003).

2   The breakthrough research on this was the work of Mario J. Molina and F. S. Rowland (1974) who found that CFCs were acting to destroy the ozone layer in Antarctica.

international community accepted these costs and continued to look to ways to further improve the two treaties.

For the problems that both CFCs and POPs created in the polar region, the role of science has been of critical importance. Without the willingness of scientists to explore the impacts that these products were having on the environment, there would have been no effort to address the problem. When scientists such as Dewailly and Rowland made their claims, they encountered both skepticism and resistance to their findings. However, other scientists were then able to build on their work until there was a general consensus that the problems did exist and action needed to be taken. It was then into this sphere that the international community then mobilized to create treaties that required action. Thus in these cases, science did lead to important acts of cooperation in the polar regions. However, there is now a case in which the opposite is happening. In the case of climate change, there has been tremendous scientific consensus on the both the causes and impact of climate change in the Arctic. However, there are few signs that there is going to be action taken equal to what was done in regards to CFCs and POPs. Instead there seems to be an increasing move to unilateral action that seems intent on seizing the expected benefits of climate change.

By the end of the 1990s and into the 2000s the scientific community began to realize that the Arctic and some parts of Antarctica were warming. At first these changes were seen as local and minor. However, as more scientists began to study this phenomenon it was soon recognized to have a magnitude of scale on a global basis.[3] Specifically, the impacts of climate change are now understood as being of a monumental nature in the polar region. However, the reaction of the international community has been very different from that to the problems caused by CFCs and POPs. Climate change and its implied assumption of greater maritime accessibility have not brought forward an international effort to create a diplomatic solution to this environmental problem. Instead the international reaction has been the polar opposite. There is no international effort to develop a polar treaty to address the problem. The polar nations are now focusing on improving on their defence capabilities and on retaining their abilities to act in a unilateral manner in the Arctic.

Science and sovereignty in the Arctic

The international community has turned its attention to the poles through the research conducted by the fourth IPY (2007–08). The number, scope and magnitude of the scientific work have been very impressive, currently generating substantial new research. Perhaps of even greater importance has been the effort to ensure

---

3   The dominant study on this issue was produced as an international multilateral multidisciplinary study lasting several years. See Arctic Climate Impact Assessment (2004).

that this research proceeds in international collaborative fashion. Great effort has been given to supporting research across national boundaries. The other element of the IPY that has generated substantial attention has been the willingness of governments to dedicate substantial resources to the process. The Government of Canada alone has already dedicated over $150 million of new funding to support research into the polar regions. Likewise the other Arctic (and non-Arctic) states have also allocated substantial resources to the various projects. There has been some speculation that the total amount of new spending on polar research may exceed a billion US dollars. There is no doubt that the fourth IPY has provided an important stimulus to science.

Driving many of these efforts is the growing awareness that climate change is fundamentally altering the polar regions. However, unlike the reaction to the issues surrounding both CFCs and POPs, the main Arctic states seem intent on using the emerging science to protect their unilateral interests rather than responding in a multilateral fashion to the cause of climate change. There has also been a parallel track of new spending in the Arctic scientific research that has not received as much publicity and may be equal or perhaps even exceed the amounts now being allocated through IPY projects. This pertains to the efforts of the five Arctic nations that may have extended continental shelves in their offshore Arctic regions. Canada, Russia, Denmark (for Greenland), Norway and the United States are all currently engaged in the mapping of their offshore Arctic regions. This is being done in the anticipation that they will then be able to extend their regions of national control.

Under the terms of the United Nations Convention on the Law of the Sea (UNCLOS) (specifically Part VI – Article 76), states that can scientifically demonstrate that they have an extension of the continental shelf off their coastline are able to claim sovereign control over the resources on the soil and subsoil beyond what is now allowed under international law. Currently all coastal states have sovereign control over the resources contained in offshore region known as the Exclusive Economic Zone (EEZ). It extends 200 nautical miles from their coastline outward. While the coastal state does not have total sovereignty over this ocean region, they have control over all economic activity except for shipping. Under Article 76 of UNCLOS, if it can be proven that there is a physical extension of the continental shelf, this area of control may be extended an additional 150 nautical miles for activity on and beneath the seabed. The purpose of this right is to give the coastal state the ability to control any offshore resource development that could occur in this region. The specific interest is in the possibility of oil and gas development.

In the case of the Arctic, it appears – but has not yet been confirmed – that a substantial amount of the Arctic Ocean may be part of the continental shelf as defined by UNCLOS. Under the terms of the Convention, a state has 10 years from the date that it ratifies the treaty to undertake and then produce the scientific data to demonstrate that it has a continental shelf. A state submits this claim to a body that was established under the terms of the Convention – the Commission on the Limits of the Continental Shelf (CLCS). The role of the Commission is then to evaluate the scientific evidence. If the Commission agrees that the state making the submission has demonstrated that it has an extended continental shelf, the state is then requested to resolve any potential overlap that it may have with its neighbors. The expectation is that potential boundary disputes will be resolved through the dispute settlement mechanisms contained within UNCLOS (Part XV). However, there has yet to be the situation in which two states have been ready to resolve any difference.

In the Arctic, the efforts of the five states that may be able to claim an extended continental shelf are being fuelled by the recognition that there may be substantial reserves of oil and gas. The United States Geological Survey has recently estimated that the Arctic may contain 30 per cent of the world's undiscovered natural gas and 13 per cent of the world's undiscovered oil (Bird et al. 2008). As a result there is what many have characterized as a developing 'resource race' in the Arctic (Borgerson 2008) as reflected in the efforts of all five coastal nations engaged in the mapping their seabed floors. While there has been some international cooperation between Canada and Denmark and more recently between Canada and the United States, it is clear that all five of the states involved have recognized the importance of allocating substantial resources to the national determination of making their claim. Thus it appears in Canada the total amount of resources dedicated to this effort alone may equal or perhaps exceed the total amount that the Canadian Government has dedicated to the entire IPY program.

Of the five Arctic states, the Russians were the first to begin to develop their claim. They submitted a claim to UNCLOS in 1999 but were requested by the CLCS to provide more data. To this effort they have also dedicated substantial resources. They also created considerable international controversy when in the process of carrying out the research, they took the opportunity to use mini-submarines to plant a Russian flag at the North Pole. Regardless of the diplomatic ramification of this act, it underlined the effort that they are making.

There is no means to currently know the total amount of resources – in terms of both finance and personnel – that is being dedicated to this effort, but it is substantial. Once all of the states have made their claims, it may be possible to get an approximate idea of the total costs, but this is unlikely. The Russian government has not publicly released the cost figures of any of their efforts, nor does it seem likely to do so. However, it appears that the total amount will equal if not exceed the effort now dedicated to IPY.

One of the developing sub-texts of the effort of the five Arctic nations to determine their Arctic continental shelf is the impact it is having on the development of international governance. As recently as May 2008, the Danish government hosted a meeting with the four other Arctic coastal states to reach an agreement on how to address any potential differences that may arise. In a declaration that was issued following the meeting, all five states agreed that there was no need for an international treaty (Illulissat Declation 2008). In other words, they wanted to ensure that they each had the maximum freedom of action within their prospective new zones. Some observers have suggested that regardless of the right that these states may enjoy in their new zones, they still need to have international agreements to govern the environmental protection of the entire region. Given the nature of the Arctic eco-system, an accident related to the development of oil and gas will affect the entire region. And yet there is a clear movement against the development of any such agreement. In this manner the 'science' that is being conducted is now leading the Arctic states to increase their unilateral actions.

Science and politics in the Arctic

Thus, it can be seen that as the 'science' warns of greater melting, the Arctic states are actually moving to increase their ability to act on their own. This is not creating a move towards greater cooperation but instead is driving states to engage on their own. So why is there a difference?

It should be obvious that part of the reason is that science does not automatically led to international cooperation. Science in the polar regions has uncovered the source and causes of many environmental threats and challenges. But what is done with that knowledge then depends on the type of interest that the states involved attache to the new information. When it is apparent that the information provides insights that show the state being threatened, strong multi-lateral action becomes a reality. Thus, a treaty limiting the production of certain chemicals was developed and accepted. However, when science demonstrates that changing conditions may in fact allow for new opportunities, states become much more inclined to act alone. They will even be willing to spend substantial resources on of their own to support scientific studies which further support those new opportunities for themselves.

This then leads to a very important question regarding the science now being supported by the IPY. What is being done to understand how politics and science mixes in the polar regions? How much of the IPY is dedicated to pursuing the questions that are best answered by the social sciences? After all, if one of the core objectives of IPY is to improve international cooperation, then supporting the research that speaks to the interaction between the findings of the physical sciences and policy must be a key element of the current IPY research program (IPY 2005).

Surprisingly and disappointingly, this is not the case. Instead of finding numerous projects that are attempting to understand how good science can be translated into good policy, IPY is supporting very few policy studies. The small numbers of IPY projects associated with improving international governance in the polar region including Project 861, 'Arctic Change: an Interdisciplinary Dialogue between the Academy, Northern Peoples and Policy Makers', principal investigator Ross Virginia Dartmouth University; Project 488, 'Pilot Project for Coordination of Northern Voice in the Development of National Policy', principal investigator Tracy Erman; and Project 227, 'The Political Economy of Northern

Development', are some of the very few that attempt to address issues of policy (see IPY 2008).

It is not clear if the small number of policy related projects represents a bias against policy studies or if there are simply few social science researchers that have been willing to apply for IPY funding. Regardless of the cause, the result is the same: few researchers have been willing to engage the larger policy issues surrounding the Arctic.

It has only been in recent years that there have been some signs that this is changing. There are now several initiatives that are now attempting to examine the larger policy issues that flow from the scientific findings of the changes to the polar regions. Some examples are the European Commission funded *Arctic Transform*;[4] the Aspen Institute's *Dialogue and Commission on Arctic Climate Change: The Shared Future*;[5] the Norwegian Defence Institute's *Geopolitics in the High North*[6] and the Canadian Arctic Resource Committee's *(CARC) 2030 North*,[7] to name but a few of the recent initiatives. What is striking about all of these initiatives is the fact that they are all new and none utilized the funding provided directly by IPY.

An additional problem is that individual states have also chosen not to fund projects that have attempted to address policy issues. The Government of Canada (2008) has funded 44 projects with a total sum of $150 million. Yet an examination of the selected projects shows a very strong inclination towards projects that are based on the natural sciences. Of the 44 projects only seven can be understood as being based on the social sciences. And of these projects, all deal with issues

---

4   This is led by four institutes: Ecologic (Germany; project lead), the Arctic Centre (Finland), the Netherlands Institute for the Law of the Sea (Netherlands), and the Heinz Center (United States) and is designed to 'promote mutual exchange among EU and US policy makers and stakeholders on policies and approaches in the Arctic in the stakeholder working groups; to provide a comparative analysis of existing policies and make recommendations with substantial buy-in as to how to strengthen co-operation between the EU and US; and to encourage dialogue and thus improve conditions for further transatlantic policy development and more effective protection of the Arctic marine environment' (Arctic Transform 2008–09).

5   The Aspen Institute (2009) is convening a commission to examine 'the adequacy of current institutional arrangements and international policies to effectively and sustainably manage new levels of commercial and economic activities in the region. The Commission and its work groups will pay particular attention to the need for greater international cooperation in promoting conservation, sustainable development, and shared responsibility'.

6   This programme proposes to develop new knowledge about the interaction of actors in the High North with a focus on the strategic and political developments in the circumpolar world (Norwegian Institute of Defense 2008).

7   This is a conference that is to review the northern science policies and practices of other circumpolar countries, and examine Canada's history of northern science policy, including recent International Polar Year initiatives (CARC 2008).

pertaining to local communities and none with larger issues of international cooperation and governance.[8]

Thus, it is possible to see a developing problem. Science by itself does not lead to better international cooperation. It may be hoped for, but it does not occur on its own. How new scientific information can be used to help improve cooperation is not well understood. Yet there does not seem to be a willingness to support the social sciences to understand this process. If the IPY is to improve international cooperation, it needs to be willing to examine how these processes work and not simply hope for the best. It may be that the findings of dramatic changes that IPY-based research uncovers encourages the development and support of the new policy oriented initiatives that are just now coming online. However, at this point it is impossible to know if that will occur. What is clear is that it does not automatically follow that science leads to international cooperation. Rather, cooperation can only proceed if the conditions that led to it are examined as closely as the conditions now leading to a physically transforming Arctic.

References

Arctic Climate Impact Assessment (2004), 'Impacts of a Warming Arctic: Arctic Climate Impact Assessment' (Cambridge: Cambridge University Press).

Arctic Transform (2008–09), 'Arctic Transform. Transatlantic Policy Options for Supporting Adaption in the Marine Environment', <http://www.arctic-transform.eu/index.html>, accessed 8 January 2009.

Aspen Institute (2009), 'The Aspen Institute Dialogue and Commission on Arctic ClimateChange',<http://www.aspeninstitute.org/site/c.huLWJeMRKpH/b.263 8873/k.E065/Dialogue_and_Commission_on_Arctic_Climate_Change.htm>, accessed 8 January 2009.

Bird, K.J., Charpentier, R.R., Gautier, D.L., Houseknecht, D.W., Klett, T.R., Pitman, J.K., Moore, T.E., Schenk, C.J., Tennyson, M.E. and Wandrey, C.J. (2008), 'Circum-Arctic Resource Appraisal; Estimates of Undiscovered Oil and Gas North of the Arctic Circle', U.S. Geological Survey Fact Sheet 2008–3049, <http://pubs.usgs.gov/fs/2008/3049/>, accessed 8 January 2009.

Borgerson, S. (2008), 'Arctic Meltdown', *Foreign Affairs* 87:2.

CARC (2008), 'Welcome to 2030 NORTH', <http://www.2030north.carc.org/index.php>, accessed 8 January 2009.

Dewailly, E. and Furgal, C. (2003), 'POPs, the Environment and Public Health', in Downie and Fenge (eds).

---

8   This author did submit a proposal to examine new forms of governance in the polar regions. The research team included leading international legal and politics experts from both Canada and Australia. However, it was not funded by the Canadian Government.

Downie, D. and Fenge, T. (eds) (2003), *Northern Lights Against POPs: Combating Toxic Threats in the Arctic* (Montreal and Kingston: McGill–Queen's University Press).

Government of Canada (2008), 'Science and Research Canadian Science and Research Projects selected for International Polar Year 2007–2008 funding from the Government of Canada', <http://www.ipy-api.gc.ca/pg_IPYAPI_050-eng.>, accessed 8 January 2009.

ICSU (2004), 'A Framework for the International Polar Year 2007–2008', ICSU IPY Planning group, <http://www.icsu.org/Gestion/img/ICSU_DOC_DOWNLOAD/562_DD_FILE_IPY_web_version.pdf>, accessed 8 January 2009.

Ilulissat Declaration (2008), Arctic Ocean Conference Ilulissat, Greenland, 27–29 May 2008, <http://www.um.dk/NR/rdonlyres/BE00B850-D278-4489-A6BE-6AE230415546/0/ ArcticOceanConference.pdf>, accessed 8 January 2009.

IPY (2005), 'Objectives IPY 2007–2008', <http://classic.ipy.org/development/objectives.htm>, accessed 8 January 2009.

IPY (2006), 'Edinburgh Antarctic Declaration on the International Polar Year 2007–2008', <http://www.ipy.org/index.php?/ipy/detail/edinburgh_antarctic_declaration_on_the_international_polar_year_2007_2008_1/>, accessed 8 January 2009.

IPY (2008), 'Expressions of Intent and Endorsed Full Proposal Databases for IPY 2007–2008 Activities Accessing the Expressions', <http://classic.ipy.org/development/eoi/index.htm>, accessed 8 January 2009.

Molina, M.J. and Rowland, F.S. (1974), 'Stratospheric Sink for Chlorofluoro-methanes: Chlorine Atomcatalysed Destruction of Ozone', Nature 249, 810–12.

Norwegian Institute of Defense (2008), 'Geopolitics in the High North', <http://www.geopoliticsnorth.org/index.php?option=com_content&view=frontpage&Itemid=1>, accessed 8 January 2009.

Stockholm Convention on Persistent Organic Pollutants (May 2001), available at <http://chm.pops.int/Portals/0/Repository/convention_text/UNEP-POPS-COP-CONVTEXT-FULL.English.PDF>, accessed 8 January 2009.

United Nations Environmental Programme Ozone Secretariat (1987), 'The 1987 Montreal Protocol on Substances that Deplete the Ozone Layer', <http://ozone.unep.org/Ratification_status/montreal_protocol.shtml>, accessed 8 January 2009.

Chapter 4

IPY Field Stations: Functions and Meanings

Urban Wråkberg

Introduction

Early scientific stations in the Arctic were mainly organized for logistical purposes. They provided a forward geographical point of departure after over-wintering for expeditions to set out for new geographical records (Wråkberg 2001; 2004b). Such activity was the main target of criticism launched against previous polar research by promoters of the first IPY 1882–83. This initial attempt at international scientific cooperation, the first of its kind in the polar regions (Barr 1985), was based on the idea that synoptic data collection undertaken by professionals at a secure, well-designed polar station would produce knowledge of superior quality and interest to science than that collected until then by travelling ship– or sledge-based polar expeditions (Baker 1982).

Today many research stations are in operation in the polar regions. Those situated on the archipelago of Svalbard/Spitsbergen in the European Arctic characteristically are often joint operations by a number of nations, or run by specific national research organizations. This is partly due to the special Spitsbergen Treaty that grants citizens of all signatory nations equal access to Svalbard within the laws established by the Norwegian government. Norway has held sovereignty over the islands since 1920. Prior to that Svalbard was a *terra nullius* with regard to national identity and was called by its Dutch name Spitsbergen. Science has long played a role as a means by which countries could show an interest in the region, making the islands and their research stations an ideal laboratory for reflections on the interplay between science, politics, industry, environmentalism and international law in the polar regions.[1]

This chapter focuses on polar research stations, their traits, and their meaning in the surrounding landscape and in international politics. Research stations will be discussed over an extensive time-frame to facilitate observations on related international political issues. While most attention will be concentrated on Svalbard, polar stations elsewhere will also be considered.

The dependency of polar science on political and economical forces operating in the far north and far south has been discussed in recent studies (see, for instance, Bravo and Sörlin 2002; Elzinga et al. 2004a; Drivenes and Jølle 2006). However,

---

1 The oldest house still standing on Svalbard is the restored station of the Swedish over-wintering team that participated in the first International Polar Year 1882–83.

the role of science and the importance of science lobbies in the formation of national agendas are often underestimated with regard to the polar regions. It can be demonstrated that even such seemingly objective parameters as the location, design and layout of polar research stations are in many ways shaped by political concerns. Polar stations are laden with cultural values and thus have proven to be useful tools for a variety of political interests and contexts. Field stations also deserve further study for the role they play in research, diplomacy, the local surroundings and the social fabric of the area, both during and after their initial period of use.

Many organizers of different IPYs have expressed pragmatic political interest in building and maintaining polar stations that is sometimes far removed from altruistic devotion to international science. The presence of a long deserted research station situated at a remote Arctic or Antarctic location and bearing a definite national identity could prove valuable for one party or another in disputes over territorial sovereignty. Thus, the first Swedish IPY station on Spitsbergen (Figure 4.1) was declared Swedish state property long after its time of initial IPY service and figured in a number of attempts to prove Swedish scientific and territorial priority on the Svalbard archipelago before international negotiations settled the issue in favour of Norway during post–First World War mediation in Paris (Wråkberg 1999b, 2003, 2004b).

Figure 4.1   Swedish IPY station in operation from 1882 to 1883 at Cape Thordsen on Svalbard (European Arctic), as it appeared in 1997. Photo: Urban Wråkberg

Sociological studies based on interviews and participant observations have been conducted on some polar research stations (Krupnik et al. 2005; see also Weiss and Gaud 2006; Weiss et al. 2007; Dozier in Bravo et al. 2007). Scientometric evaluations are regular parts of reviews made of contemporary research facilities. However, activities in the polar regions, as well as their cost, have not only been motivated in many different ways, but also funded by an odd array of supporters. Scientific productivity may be only one of many desired functions of a polar research institution (see Bravo et al. 2007), nor can these diverse activities often be separated: many scientists and scholars specializing in the polar regions have also been committed lobbyists, environmentalists, industrial promoters and, in some cases, legislators of national policy on polar and sub-polar regions.

Today most IPY research teams are multinational in composition. However, this was not the general pattern during the first two IPYs. Internationalism has several sides; it can also help push national commitment through the element of competition. Ever since the first IPY this fact has been successfully used to argue for national funding by reference to the participation and ambitious scale of other national research commitments, as manifested by the number, size and location of their polar field stations (see Elzinga 2004b on internationalism and its obstacles in polar research).

The IPY has always been inspired by a mixture of ideas and aspirations, where urgent scientific issues, new technology and political and territorial interests converge. There is a continuing applicability of polar science and expertise in a military context and a growing one in resource detection, monitoring and utilization. For example, some scientific issues and advanced marine technology have new and crucial relevance to certain issues in the international law of the polar regions. This is demonstrated by the current problem of determining the boundary between the abyssal ocean floor and the continental shelf of northern coastal states, as their rights to the subsea bottom resources of the Arctic Ocean have been tied to the extent of their shelves by the 1982 United Nations convention on the Law of the Sea (UNCLOS) (on the relation between IPY, polar science and military concerns see Doel 2003; also see Huebert in this volume on international law; and Jacobsson 2004 on Swedish foreign policy and polar research during the last century).

Sciences with applications in the polar regions have always suffered from a relative lack of Arctic and Antarctic data due to the fact that research in these regions of the Earth is more risky, demanding and costly than elsewhere. The high north and Antarctica have also long been seen as the last terrestrial expanses of pristine wilderness. In the days before the poles were reached, the unknown central polar region was laden with mystery. Theories circulated such as that of the hollow earth, the paleocrystic ice-desert and the thought that the central Arctic, behind an outer ice rim, was an open polar sea – which had many adherents in the 1860s – to mention just a few obsolete scientific ideas.

The ring-shaped Arctic network of IPY stations during the early polar years formed a colonial frontier encircling untamed nature with stations that were outposts or forts. The Cold War DEW (Distant Early Warning) Line and clandestine drifting polar stations were vital surveillance components forming the northern front of global defence systems. A related imagery also inhabits contemporary alarmist science, which envisions polar stations and their observatories as a first line of lookouts equipped with early warning systems designed to detect threats to humanity in the form of extreme weather, lethal enrichment of pollutants in food chains, ozone holes, climate change – as if nature will finally strike back at humanity from its last stronghold in the polar wilderness. The polar station is an entity designed, constructed, and utilized in various ways within a broad context of notions, scientific ideas, media reports and international and regional politics (Figure 4.2).

Figure 4.2   NATO surveillance radar in Vardø on northeastern-most coast of Norway. Photo: Urban Wråkberg

The research station as a social institution

A field station is one of many locations at which scientific knowledge is produced. It is related to other institutions of scientific interaction, research and re-production such as universities, institutes, academies, observatories, laboratories, testing ranges, excavation sites, field camps, research vessels and expeditions. Over time, field observatories and laboratories have become integrated parts of more ambitious research stations. One of a station's original purposes was to provide shelter and a stable environment in which to house standardized instruments for precision measurements of meteorological and geomagnetic parameters (Figure 4.3). This formed a central argument in the rhetoric of the early IPYs: sensitive yet delicate scientific instruments were specially developed to be transportable to stations in the field by ship and manpower. A controlled environment was needed in which they could be installed and calibrated in order to be operational. Travelling expeditions on sledges and ships could never provide the protected circumstances essential for the production of precision data, just as an extended series of observations were not within their capabilities either as they normally did not reside in any one well-defined geographical position for very long. A circumpolar network composed of an array of stations situated at some distance from the pole could produce more data of interest to science for the equivalent cost as that of launching a single, dramatic, resource-consuming and risky expedition towards the pole. Nevertheless this classic IPY line of reasoning has known many modifications (Breitfuss 1930).

The idea of landing a well-equipped observatory on a remote shore somewhere in the far north or south, then operating it as one would do 'back home' on a university campus, was never realized during the first polar years, and it remains an unfulfilled goal in much contemporary polar research. But to overlook the dependency and interrelation of a polar station with its surroundings is misleading and potentially hazardous, as has been proven on more than one

Figure 4.3   **Observation hut at IPY1 station, Spitsbergen, photographed in 1883. Instrument visible beneath opened roof is a field version of an astronomical transit. By coordinating passage of stars across the celestial meridian with local time, precise measurement of the observatory's geographical position was possible. When not stated otherwise the illustrations in this book are published by permission of the Royal Swedish Academy of Sciences**

isolated a way as possible into the field, all successful polar explorers of the past and many contemporary station leaders have recognized the value of good local collaboration. Indigenous experts have not only provided crucial operational and survival information, but have pointed the way toward new lines of research on the surrounding environment and culture. (The complexities of the dialogue between Western and traditional knowledge past and present are discussed by Bravo 1996, Ellis 2005 and by Shadian in this volume.) Field stations relate to their locations on both a micro socioeconomic level as well as an intermediary level of landscape and region. In the case of the IPY, a third global level of coordination, namely that of its internationally agreed programme of standardized and synchronized scientific observations, is of particular importance.

Another reason to avoid an unreflected internalist view of IPY research is that the international scientific community has never been able to articulate either a single agenda or an agreed upon priority for the IPY. Despite the arguments of some of its proponents over time, the IPY never was the obvious logical step forward in order to transform national polar research into an international endeavour. The original IPY programme was always more attractive to certain scientists in the fields of meteorology, climatology, geomagnetism and atmospheric physics, who needed a stationary observatory, than to others (see Nilsson in this volume). Cartographers, geologists, ethnographers, oceanographers and botanists historically have held a stronger position than the former in polar research and they all required mobility to carry out their field-work.

In order to make a map, cartographers must travel extensively, field geologists need to sample minerals in many locations, ethnographers have to interview people wherever they may live and oceanographers routinely measure water temperatures, salinity and currents not only at points on the surface of oceans but also at regular intervals down to the bottom of the sea. If the latter polar researchers were to need a polar station, it would only be as a base camp. Thus the IPY and its network of field stations was never a united enterprise for all scientists not even if we accept to misconceive science as some kind of closed autonomous but global business. (For scientific disputes regarding the early IPYs, see Wijkander 1880, 2; Hellwald 1881, 769 f; Neumayer 1901, 205–24; Drygalski 1898; Levere 1993, 313 ff.) As will be discussed further in the following sections, when IPYs were eventually launched, they were based on special funding, often motivated more by geopolitical interests than by any international scientific unanimity on the urgent need for more knowledge of the polar regions.

The importance of physically controlling the inside of a research station in the interest of reproducing precision measurements is critically related to the need for exerting social control over the same space, and of providing resident observers with comfortable accommodations. The difficulty – perhaps impossibility – of finding one person, or even a small team, combining excellence in scientific research, field leadership, talent for logistical planning, and physical fitness, has been clear since the days of the grand race towards the poles. A field station presents an alternative arrangement that addresses the problem of having to create such an ideal team by providing a group of less daring scientists with a comfortable enough environment to enable them to function and do valuable work, although at an extreme location (Levere 1988, 233–36; Wijkander 1880).

The control of the inside of the field station also means limiting access to only certified personnel who have been disciplined in observational practices. But as in the case of travelling expeditions, field stations have also served as clearinghouses of knowledge gathered from amateur observers, laymen and indigenous informants encountered in the field. The social role of the field scientist includes a position for evaluating the credibility and scientific value of such outside findings, as well as the authority to introduce them (in a shape more or less refined) into the international body of accepted scientific data. The result of such attempts depends, in turn, on the professional standing of the scientist involved. (Wråkberg 2002, 159, 173; Powell 2002, 263; cf. McCook 1996).

Some people see the scientific management of winterings as a unique endeavour. Although other forms of organization, such as naval expeditions or ventures based on common business interests, have proven successful on occasion, many found the combination of military and scientific cultures to be particularly disastrous during expeditions or winterings. Early on, the polar field station and research vessel found their identity as distinct, scientifically guided undertakings (Figure 4.4) (Eliasson 2001; Rozwadowski 1996; Capelotti forthcoming).

It would be outside the scope of this essay to examine the internal culture of the polar field station in detail. Many odd tales have been told that seem to mirror the extreme isolation and the considerable costs and risks involved in that life. Some have described heterotopias where collegial bonding of life-long importance takes place; peculiar rites that seem to repeat themselves during winterings, such as improvised lonesome Christmas or birthday parties; the social and emotional importance of caring for draught animals like dogs as pets; the passing of the long winter by writing, staging and collectively performing 'polar plays', cross-dressing happenings, odd field traditions prescribing that the cook should always be the youngest member of the team, the reluctance to allow women to participate in winterings, or to include in official reports that they actually did.

Other observations reported from polar stations regard the practice of leadership: some leaders tried to fight recurring conflicts by issuing a ban on talking before midday, while others mastered Arctic lethargy successfully by rigorous routines and elaborate work schedules for everyone based on a silent agreement not to ask about the usefulness of the work.

Automatic weather observatories and remotely operated download centres of satellite data at contemporary polar stations seem to make all this management of human behaviour redundant, but any Arctic or Antarctic station is still an extreme place in most social and logistical regards, and the seemingly neutral matter of anyone being present in a research station at a certain location can often be a kind of territorial investment with a variety of political meanings.

Figure 4.4   **Over-wintering team at Swedish IPY1 station on Spitsbergen in 1883**

The kinds and uses of field stations

Judged by public arguments put forward on different occasions for erecting field stations, they have been constructed for purposes as various as depots, base or rescue camps and centres for military or environmental monitoring. Among the tasks of the early IPY observatories were the recording of annual meteorological data and synoptic measurements of auroras.

The design of any field station is not only conditioned by the overt purposes listed above, but also embodies many other considerations. First of all it is dependent on the kind of competence that guided its construction, the size of the funding that has underwritten it, and the extent to which its structure was planned or was the result of circumstance. Models for its design are sometimes found in other Western buildings in its periphery: colonial trading stations, missionary settlements and police outposts. But the experiences embodied in houses and camps of indigenous people have seldom been utilized in the design of scientific field stations.

A popular idea in nineteenth and early twentieth century was to apply presumed national styles from the native country of the field researchers in the buildings of their polar stations (Figures 4.5 and 4.6). This was to somehow indicate the predestination, or at least suitability, of the explorer's culture to colonize and expand into the polar regions (compare the writings of Vilhjalmur Stefansson, a late exponent of colonialism in the polar regions, in Stefansson 1921, 1922). By the style of the building a distinct expression of national presence can be made in a non-verbal yet ostentatious way. Scandinavian stations, for example, sometimes carried elements of old Norse buildings, or copied the outer pattern of rural northern farms at home. Later examples include the choice of a typical national colour in the exterior painting. The size of a station, and the apparent permanence of the materials and construction used, can demonstrate the strength of political commitment and scientific ambitions – at least at the time when it was built.

Some abandoned polar stations have been left relatively intact due to their locations in remote regions that were sparsely populated until recent times. Building materials have often been well-preserved by the climate, with wood in some cases even freeze-dried. An older polar field station may still be in use, perhaps in a substantially altered or expanded form; or it could no longer be put to its original use and instead be transformed into a touristic or archaeological object. With the passage of time, the condition of the field station and the extent to which it has been protected, restored or used (for example, as a cultural monument) becomes a significant indicator of the influence of various social forces on the site.

The physical state of preservation of a research station may vary considerably. Several of these were not built on stable ground. Solid rock is not to be found at most locations in Antarctica, which means that anything constructed will eventually sink into the ice, be covered in snow and layered into the ever-flowing ice sheet (Figure 4.7), especially when out of service and no longer maintained. In the Arctic, temporary stations drifting with the sea ice have been used systematically since the

Figure 4.5   Hoisting flag at northern research station of Swedish-Russian Arc-of-the-Meridian measurement programme in Spitsbergen, 1899–1900. This house of typical Swedish national style was later completely demolished

1930s for both military and civilian purposes. Several abandoned Antarctic camps, like the US IGY station on the South Pole, have sunk beneath the surface and been crushed in a deep-frozen state. Their availability for future archaeological or environmental impact studies is thus rather uncertain (West 2008; on management issues of the cultural heritage of the polar regions, see Barr and Chaplin 2005).

Pre-fabricated buildings have long been of special interest in the polar regions. Several pioneering designs were tested as polar field stations. However, such structures made of timber and planks required the transport of both carpenters and building materials, as the sites chosen were normally situated beyond the tree-line. Local building materials used by aboriginal and resident hunters, such as stone, driftwood, earth, turf and snow, were unfamiliar to Western designers and seldom used. This results in the often-repeated 'neutral' material explanation as to why

Figure 4.6   Russian research station in Hornsund, southern Spitsbergen, in 1900. Design by Aleksander Bunge, based on his extensive field experience and local Siberian customs. Low, spacious building minimized wind exposure while storage rooms placed around bedrooms and central living area provided insulation

Figure 4.7   Norwegian-British-Swedish Antarctic research station 'Maudheim', almost buried in snow during first winter of service in 1950

deserted buildings have been demolished by being reused as a source of materials for other buildings, much as later Romans stripped the Colosseum, or simply as a supply of firewood.

While sometimes motivated by emergency, and in many cases facilitated by the lack of surveillance and law-enforcement of anyone's rights of property in remote polar regions, the reuse or pillage of deserted stations is also largely the result of public policies in the area. The active destruction of buildings might also be inspired by various ideologies or the outcome of business competition. Passive destruction by what 'nature takes back' through the process of erosion and corrosion is nowadays primarily the result of land management goals and priorities (Hughes 2000). Many modern buildings erected for scientific and other purposes have been dismantled or destroyed in areas which have been accorded the status of nature preserves before or after the construction of the station. Here again we encounter policies that may not only be determined by concern for nature, but also by issues of sovereignty and the interest to exclude newcomers and late starters from Arctic or Antarctic research.

The deserted field station, in its various states of preservation or decay (Figures 4.8 and 4.9), mirrors policies of removal or conservation and their value-laden application. It also reflects the competence and self-interest that enters into matters of preservation: attitudes on reuse, regarding clearing and cleaning, and the protection of nature versus cultural heritage. It is revealing to analyze the policies and the implementation of heritage regulations as applied over time in various territories, and ask whether any professional expertise involved is scientific, archaeological, historical or political. One may similarly question the extent of international orientation and awareness. Natural or cultural heritage has seldom been viewed as something relevant (and in some sense belonging) to the whole of humanity (Appiah 2006; Merryman 1986); instead, the prevailing notion seems to be that national territory may be handled according to the principles of management applied in 'your own backyard'. In the polar regions this often means that anything there of general value is 'owned' by an elite in a distant capital with the powers to define environmental and national interests, and to control all major resources based on less-than-transparent policies and implementation strategies.

It is clear from recent Arctic events that the research station keeps, if not strengthens, its usability as a soft diplomatic tool by which to confirm national commitment to sovereignty of polar territories. This is based on the perceived need of using polar territories so as not to risk losing them – a line of reasoning that could be contested based on international law. However, the prospects of increased international shipping through the Northwest Passage, and that of a global scramble for the natural resources of the seabed of the Arctic Ocean, has been regarded as a scenario that has to be met by a concerted Canadian response.

A series of measures to this end was presented in 2007 by the Canadian Prime Minister Stephen Harper as part of his government's new strategy for the North. The various policies launched in this package all have the explicit additional aim to affirm Canadian presence and sovereignty in the Canadian Arctic. This includes

Figures 4.8   Astronomical observatory of Arc-of-the-Meridian measurement
programme in Spitsbergen, photographed from beach in 1898

Figure 4.9   Scant remains of same building as seen in 1998, looking towards
bay

measures such as strengthening the ability of the northern territorial governments to deliver housing for First Nations and Inuit. Expanded military and coast guard presence in the High Arctic are among the well-known means in this context, as are improved surveillance capacity and the reinforcement of the Canadian Arctic Rangers. But also stepping up environmental activities by increasing the number and extension of National Park Reserves in the North were mentioned by Harper in his opening speech for the Canadian parliament in November 2007, as was enhancing northern research by engaging in the IPY. Harper pointed out that 'These research activities will help confirm our unassailable ownership of the Arctic Archipelago and the waters around them, including the Northwest Passage, along with the resources that lie beneath the land, sea and ice'. In this speech Harper also presented the idea to establish of a world-class research station to be located in the Arctic itself (Harper 2007).

The present Canadian government's Northern Strategy is to deliver a further developed version of the ideas on polar research presented by the prime minister in 2007, in what is called the Canadian Arctic Research Initiative. Favourable comments on this were immediately presented from both First Nation Canadians' representatives as well as the Canadian research community (CBC News 2007). It is worth noticing that the commitment of other nations to polar field research is interpreted as putting a pressure on Canada to assign further resources to this end; thus the double action of international research provided by IPY lobbyists in offering both collaboration and competition. In an effort to contribute to the development of the Canadian Arctic Research Initiative, the Council of Canadian Academies, a federally funded science advisory body, in 2008 set up an international panel of experts to review the options for the initiative outlined so far in Canada. The panel was composed of research leaders and senior scientists from Europe, Britain, the US and Canada (International Expert Panel on Science Priorities for the Canadian Arctic Research Initiative 2008, 2).

The panel's report gives no specifics on facilities or costs, but it does recommend a much-expanded vision of what Harper initially launched as a new cutting-edge Canadian Arctic Research Station. The panel stresses that a more robust Arctic science program is essential and that Canada may even hinder the development of global Arctic knowledge unless it is fully engaged. The panel recommends that Canada position itself as steward of its remarkable global resources, to build on the results of the International Polar Year and to reassert Canada's place in the international polar research organizations, recommending that continual funding is provided to assure continuity and follow-up on the research initiatives of the IPY4 (International Expert Panel on Science Priorities for the Canadian Arctic Research Initiative 2008).

Among the many well-founded pieces of advice from the expert panel, the report interestingly provides good arguments, not only for one polar field station but for a large permanent network of them. It endeavours to combine interests together across a wide and powerful range of sectors of Canadian society, and thus to connect the political forces of the environmental, industrial and national security spheres to the less powerful research community. Being the advice of an international panel this also seems to represent a happy confluence of the interests of the Canadian scientific community with that of the international scientific enterprise – or at least of those members of both unities with a truly cosmopolitan mind.

The polar station in its material and symbolic field

In trying to make sense of the exact location at which a polar station is ultimately situated, one often resorts to public statements made by the agents engaged in establishing them. The standard pattern of these narratives generally includes a disavowal of responsibilities in which the high ambitions of the original plans are contrasted with modifications attributed to bad weather and adverse ice conditions in the field during the crucial time window in which the site needed to be chosen and debarkation undertaken. This may or may not represent the major part of the truth.

Another compelling line of explanation for the precise choice of location of polar stations is to refer to a few crucial assets of any modern Western base location: the coexistence of a natural harbour or landing strip with a nearby source of fresh water. There is also the material logic of placing a station in a position which is sheltered from the prevailing winds, or elevated to avoid getting buried in snowdrift where high levels of precipitation are expected (Figure 4.10).

In addition to such factors, an evaluation of the character of the landscape is necessary. Will the sight be conspicuous to visitors or travellers passing by? Will it be a suitable vantage point for observing and monitoring the surroundings? Is a concealed location designed to keep the presence of its inhabitants secret

Figure 4.10   Swedish-Argentinean Antarctic field station 'Snow Hill', in use from 1901 to 1903. Prominent elevation exposed to wind kept it free from drifting snow

desirable? The latter two characteristics often apply to scientific field stations designed to operate in times of war. Those stations intended to display a grandiose commitment, proclaim national identity (by architectural or other traits), and embody territorial interests, are located in prominent, widely visible locations.

Some of those in charge of establishing the scientific stations of the second IPY, which occurred during the world economic recession of 1932, had to resort to the practice of reusing discarded structures and buildings in the far north – and were happy to do so, considering that German national participation was cancelled altogether. The Swedish IPY contingent during the second IPY on Svalbard, for instance, housed their research teams on geomagnetism and auroras in the abandoned mining camp of Spitsbergen Swedish Coal Ltd. that went bankrupt in 1934 (Lüdecke 1993, Swedish Polar Year Expedition 1939; on the industrial heritage of coal mining on Svalbard, see Avango 2005).

By contrast, the impressive Danish station built during the second IPY on Ella Island, Eastern Greenland (Figure 4.11), was specially designed by the experienced polar research director Lauge Koch. It is very well preserved to this day, with a well-ordered interior that includes individual sleeping quarters, a radio room and many other modern conveniences. It was achieved by combining a proven design, based on extensive polar experience, with ample funding. The station was part of a substantial Danish effort involving polar research and general activity on Eastern Greenland during the second IPY, whose field campaign in 1932 involved 90 people, five vessels, and aircraft. It was propelled by the geopolitical conflict going on at the time with Norway, which then formally and publicly contested

**Figure 4.11   Danish IPY2 station on Ella Island, Eastern Greenland, in 2003.
Photo: Urban Wråkberg**

Danish sovereignty in the area. The dispute was settled in favour of Denmark in 1933 by the Permanent Court of International Justice in The Hague (Ries 2002, 206 f.; see Fogelson 1992, Nilsson 1978).

Sometimes revealing differences can be detected by not only comparing actual polar enterprises, but by contrasting the published accounts of a deserted field station, usually focused on positive achievements, with the disorder and makeshift solutions still traceable in the remains of it in the field. An example on this kind of discrepancy that has been demonstrated by archaeological research is the launching station on Danskøya, NW Svalbard, of Walter Wellman's dirigible, which he planned to fly over the North Pole in 1909 (Capelotti 1994 and 1999). The field remains of many old polar camps and stations display the fact that they were built or abandoned in great haste.

It is instructive to consider polar stations as located in a symbolic field, understood on a collective level. A component of this is the heritage projected on the landscape by the region's cartography. It becomes apparent in the act of contesting or accommodating a specific declaration of hegemony over the land by using a chosen map as authoritative, but also more subtly in regarding certain older maps and gazetteers as more valid than others for establishing reference points for further cartographic work. The most obvious components of maps that support or undermine identity are geographical names. The choice and linguistic forms of names on those maps that are in widespread use can lend support to national claims of ownership, or simply confirm priority in the cartographic documentation of the region (Wråkberg 2002, 2004a, 136–39).

Certain conventions of geographical nomenclature employed in the polar regions bestow honour upon an individual, institutional, ethnic or national heritage by forming place-names from names of persons or organizations (see US Board of Geographical Names 2006). Such names have commemorated the cartographer's scientific colleagues, sponsors, royalty, prominent persons, as well as expedition vessels. Sometimes this involves the transfer of geographical names from the cartographer's home country. Several cartographers have tried to pre-empt competition by using translations of native or older names originally given in other languages. Some have attempted to establish the 'neutral' principle of admitting only descriptive terms from nature as geographical names. However, this suppression of historical and social aspects in naming is not neutral as it often effaces earlier place-names and clouds the identity of prior and contemporary stakeholders in the territory.

Revisions have sometimes been proposed that favours terms current among the most numerous or frequent 'users' of the toponomy. Regarding the Arctic, recourse to traditional names and proposals aimed at re-introducing such names are often motivated by the principle of chronological priority and by the necessity of recognizing indigenous rights and identity (Inuit Heritage Trust 2002). However, advocating a return to traditional names often stirs up complex political debates as to whose home the land really is. Indigenous name proposals themselves are subject to controversy over issues of ethnic details, including kinship, genus, linguistic variations and the kind of social status needed to be included in the toponymic process at all – problems which have their counterpart in the naming of public places in any social context. There remains in many settings a communications problem with profound ethical issue regarding the choice and practicality of employing certain place-names. Consider, for example, the difficulty of conducting rescue operations in a region where many of the place-names are virtually unpronounceable for a majority of those living and working there, including professionals within the transport and security sectors.

Revisions of toponomies are seldom easy to undertake and hardly neutral. Among the many things at stake here are leadership in science and the priority in local cartography. These would be dwarfed in most instances by ethnic concerns, regional interests and questions of national sovereignty. The IPYs as well as the Scientific Committee on Antarctic Research (SCAR) have all engaged in establishing and regulating the toponomy of polar regions. In recent years SCAR has compiled an authoritative inventory of the existing (and often conflicting) place-names in the southern polar region. However, its attempt to establish common principles to overcome the massive problems of the toponomy of Antarctica has thus far not met with success (Antarctic Spatial Data Infrastructure 2006; Cervellati et al. 2000). This tends to confirm the abiding symbolic significance of the landscapes in which all new and old IPY field stations are situated.

The emblematic meaning of a polar station in its surroundings differs for various visitors to the site, and changes over time. Successive parties are often socially quite diverse, and may include military personnel, native trappers, scientists, environmentalists, industrialists, public administrators, politicians, travel agency operators and tourists. To some visitors the field research station may signify the spread of civilization, while to others it represents an intrusion in their homeland. The understanding of a place is influenced by a person's experience, previous knowledge and by considering the comments and explanations of others – for better or worse – when on site in the field.

A recent example of the active use of the IPY heritage may illustrate how interests of contemporary field research sometimes can motivate measures to uphold its legacy. (The case is described based on participant observation.) It involved the deserted IGY polar research station of Kinnvika on Nordaustlandet, NE Svalbard (Figure 4.12). Kinnvika was originally built for over-wintering teams of scientists and technicians from Finland, Sweden and Switzerland who participated in the third IPY (mainly called the IGY). The station was no longer used after the end of the IGY, but its buildings have been kept closed and in good condition by the office of the Governor of Svalbard (*Sysselmannen på Svalbard*), the supreme Norwegian authority on the islands. The interior of some of the houses contains furniture and items in much the same state they were left in fifty years ago. The experience of entering the premises has been likened to going inside a time-capsule by people who have acquired the permission of the Governor to visit there.

Figure 4.12   One of the houses of the IGY 1957–59 polar station Kinnvika, Northern Svalbard, as seen in 2005. Photo: Urban Wråkberg

Beginning in 2005, a Scandinavian-Polish group of scientists and scholars lobbied to reopen this base in an effort to promote their chances to compete for IPY 2007–08 funding. Although they were soon informed that the environmental and cultural heritage regulations of Svalbard would not permit this, the publicity created by the lobbyists started to build support for a multidisciplinary IPY 2007–08 research programme on northeastern Svalbard. The popular appeal of the project of re-opening an old forgotten polar station, mysteriously preserved in the far north by the cold, proved considerable. In 2005 the lobbyists made a visit to the site, accompanied by journalists and a TV crew, and this produced a good deal of media exposure that contributed to the revival of Arctic research in Finland, in addition to its more established Antarctic counterpart. It also helped to broaden the scope of IPY research in Sweden in a more multidisciplinary direction.

Thus, this group of scientists and scholars did manage to launch a cluster of research projects in time for the IPY 2007–08 that got endorsement as an official IPY team with funding and some logistic support from the national polar research systems of Finland, Sweden and Poland. The programme was entitled 'Change and Variability of Arctic Systems Nordaustlandet, Svalbard: "Kinnvika"'. Its field research focused on issues of climate change and multidisciplinary cultural studies (Pohjola 2006; Kinnvika 2008). The old Kinnvika station's historical novelty to most Scandinavians, its fair state of preservation, and the positive meanings of it successfully shaped by the Kinnvika lobbyists turned it into a symbol nostalgically calling new generations of scientists into active polar field service. In this way it furthered institutional and individual interests of the lobbyists as they enacted old clichés of national traditions and research legacies in the media coverage produced, that are really without meaning in Scandinavia today, at least to most minds privileged to make decisions on the allocation of the meagre resources of contemporary Swedish polar research.

Field stations, as most sites of human interaction with polar nature, have malleable meanings, but some are quite rigid at any given moment, as to what they signify to a wider audience. Such inflexibility could result from interests exerted by influential stakeholders. When large investments have been made in certain meanings, re-interpretations will be refuted. The importance of major symbolic constructs goes beyond scholarship, literature and art, and into the realm of media, politics and international relations. There is a number of well-know polar dramas and their relevant field sites that have always been laden with competing interests, constructed meanings and political uses, such as the lost John Franklin expedition that sailed north to navigate the Northwest Passage in 1845, Robert Peary's disputed attainment of the North Pole in 1909, the crash landing of Umberto Nobile's airship *Italia* in 1928 and the *Kursk*-disaster in the Barents Sea in August 2000 (Wråkberg 2001).

The Swedish station of the first IPY was once used for a forced wintering by a party of young fishermen, all of whom unfortunately perished from scurvy or food poisoning. As a result, many visitors, less informed or impressed by international polar year science, perceive the site as signifying nothing but tragedy. The same holds true for many observers with regards to the American station at Fort Conger, Ellesmere Island, Canada, after Greely and his men's unsuccessful winterings there during the first IPY.

Summing up: The geopolitical meanings of polar stations

Envisioning polar regions as a land of promise for science and a potential source of wealth for the global economy continues to inspire the post-colonial world. In the Western conception of the polar wilderness, basic and applied science retains their political and economic position as the most efficient agents to unlock the polar regions and the dependable stewards of its use. Facilitated in recent times by the assistance of the social sciences, indigenous Arctic groups have entered into the expanding circle of stakeholders who find themselves summoned to guide, evaluate and control polar research. Traditional interests thus seem possible to handle in Arctic knowledge-production in ways attractive to almost everyone (Korsmo and Graham 2002). Economic geography, mineralogical prospecting, fishery science, regional development studies and other branches of research that take stock of natural and societal resources are an integral part of polar science as it continues to explore for and monitor finite and renewable natural resources. Thus, polar research is a strategic political instrument, of which the scientific field station remains an important component, with the added political value of being a signifier of commitment, territorial interest and local competence.

Although non-textual information, material sites and landscapes are seldom used as sources in science and technology studies (S&TS) or in political science, if one contrasts verbal accounts of IPY field stations with on-site observations made on field visits to the site; it may demonstrate aberrations from official aims or contextual affinity with phenomena that are not part of IPY self-consciousness. Big science promoters, such as those supporting the IPYs have had to make their activities and goals attractive to the broader public and to major decision makers outside the walls of the university. Internationalism seldom outweighs economic, political and national interests. The local, regional and national character and value of IPY science is thus often underestimated.

Field stations have proven to be flexible enough to be put to uses none of their original builders could have foreseen (Figure 4.13). Building and operating IPY stations have been distinctly national undertakings; few of the first IPY teams were international in composition. The design of station buildings, their layout and location – all demonstrate various national rather than one international character. At the same time territorial concerns have been prevalent in many early IPY programmes. Adding to the complexity of the IPY quite a few of the sciences were not really furthered by its initial station-based methodology, and the social sciences were only officially endorsed as part of the latest IPY.

Figure 4.13 **German WWII meteorological field station 'Haudegen', hidden on remote Rijpfjorden in northeastern Svalbard, as seen in 2001. Never discovered by Allies, the station gathered and transmitted encrypted strategic weather information and engaged in ambitious scientific research. Its commander, Dr. Wilhelm Dege, was among last German officers to peacefully surrender in September 1945. Photo: Tyrone Martinsson**

Several choices of sites in the first and following IPYs were determined by more or less official national 'zones of interest'. This was a consequence of the colonial world order and its division of the globe into different 'interest spheres' by the industrialized nations. Today, globalization and the regime under the Antarctic Treaty blur the picture of such zones and sectors, as does the acknowledgement of cross-border interests and communities of indigenous people. This may be a sign of change in the general direction of polar research towards identifying and pursuing common human, economic and environmental goals.

References

Antarctic Spatial Data Infrastructure (2006), 'SCAR Composite Gazetteer of Antarctica', Scientific Committee on Antarctic Research [website] (updated 20 July 2006), <http://www.antsdi.scar.org/placenames>, accessed 24 August 2008.

Appiah, K.A. (2006), 'Whose Culture Is It?', *New York Review of Books* 53:2, 38–41.

Avango, D. and Lundström, B. (eds) (2003), *Arbetets avtryck: Perspektiv på Industriminnesforskning* (Stockholm: Brutus Östlings förlag).

—— (2005), *Sveagruvan: Svensk gruvhantering mellan industri, diplomati och geovetenskap 1910-1934* (Stockholm: Jernkontoret).

Baker, F.W.G. (1982), 'The First International Polar Year 1882–83', *Polar Record* 21, 275–85.

Barr, S. and Chaplin, P. (2005), *Cultural Heritage in the Arctic and Antarctic Regions*, Monuments and Sites: new series no. 8 (Oslo: International Council on Monuments and Sites, International Polar Heritage Committee).

Barr, W. (1985), *The Expeditions of the First International Polar Year 1882–83*, The Arctic Institute of North America Technical Paper no. 29 (Montreal: Arctic Institute of North America).

Beretta, M. and Grandin, K. (eds) (2001), *A Galvanized Network: Italian–Swedish Scientific Relations from Galvani to Nobel*, Bidrag till Kungl. Svenska Vetenskapsakademiens Historia no. 32 (Stockholm: Royal Swedish Academy of Sciences).

Bravo, M. (1996), *The Accuracy of Ethnoscience: A Study of Inuit Cartography and Cross-Cultural Commensurability*, Manchester Monographs in Social Anthropology no. 2 (Manchester: Manchester University).

—— and Sörlin, S. (eds) (2002), *Narrating the Arctic: Collective Memory, Science, and the Nordic Nations, 1800–1940* (Canton, Mass.: Science History Publications).

—— et al. (2007), 'Polar Field Stations and IPY History: Culture, Heritage, Governance, 1882–Present' [website of officially endorsed IPY 2007–2009 research programme no. 686] <http://museum.archanth.cam.ac.uk/fieldstation/>, accessed 7 August 2008.

Breitfuss, L. (1930), 'Das Internationale Polarjahr Einst und Jetzt: Rückblick und Ausblick', *Arktis: Vierteljahrsschrift der internationalen Gesellschaft zur Erforschung der Arktis mit Luftfahrzeugen* 3, 14–30.

Capelotti, P.J. (1994), 'A Preliminary Archaeological Survey of Camp Wellman at Virgohamn, Danskøya, Svalbard', *Polar Record* 30, 265–76.

—— (1999), 'Virgohamna and the Archaeology of Failure', in Wråkberg (ed.).

—— (2009), 'Further to the Death of Sigurd B. Myhre at Camp Abruzzi, Rudolf Island, Franz Josef Land, 16 May 1904', *Polar Research* 28(2).

CBC News (2008), 'Throne Speech's Northern Focus "A Good Start": Inuit Leader', Canadian Broadcasting Corporation News 17 October 2007 [website] <http://www.cbc.ca/canada/newfoundland-labrador/story/2007/10/17/north-speech.html>, accessed 16 December 2008.

Cervellati, R. et al. (2000), 'A Composite Gazetteer of Antarctica', *SCAR Bulletin* 138, <http://www.scar.org/publications/bulletins/138/a.html>, accessed 24 August 2008.

Doel, R.E. (2003), 'Constituting the Postwar Earth Sciences: The Military's Influence on the Environmental Sciences in the USA After 1945', *Social Studies of Science* 33, 635–66.

Drivenes, E-A. and Jølle, H.D. (eds) (2006), *Into the Ice: The History of Norway and the Polar Regions* (Oslo: Gyldendal Akademisk).

Drygalski, E. von (1898), 'Die Aufgaben der Forschung am Nordpol und Südpol', *Geographische Zeitschrift* 4, 121–33.

Eliasson, P. (2001), 'Arktisk leda', in Eskilsson and Fazlhashemi (eds).

Ellis, S.C. (2005), 'Meaningful Consideration? A Review of Traditional Knowledge in Environmental Decision Making', *Arctic* 58 March, 66–77.

Elzinga, A. et al. (eds) (2004a), *Antarctic Challenges: Historical and Current Perspectives on Otto Nordenskjöld's Antarctic Expedition 1901–1903* (Göteborg: Royal Society of Arts and Sciences).

—— (2004b), 'Otto Nordenskjöld's Quest to Internationalize South-Polar Research', in Elzinga et al. (eds).

Eskilsson, L. and Fazlhashemi, M. (eds) (2001), *Reseberättelser: Idéhistoriska resor i sociala och geografiska ru* (Stockholm: Carlsson).

Fogelson, N. (1992), *Arctic Exploration and International Relations 1900–1932* (Fairbanks: University of Alaska Press).

Harper, S. (2007), 'Strong Leadership: A Better Canada, Prime Minister Stephen Harper addresses the House of Commons in a Reply to the Speech from the Throne, 17 October 2007', Office of the Prime Minister of Canada [website] <http://www.pm.gc.ca/eng/media.asp?id=1859>, accessed 16 December 2008.

Hellwald, F. von (1881), *I höga norden eller: Nordpolsforskningarna från äldsta till närvarande tider* (Stockholm: Palmquist).

Hughes, J. (2000), 'Ten Myths About the Preservation of Historic Sites in Antarctica and Some Implications for Mawson's Huts at Cape Denison', *Polar Record* 36, 117–30.

International Expert Panel on Science Priorities for the Canadian Arctic Research Initiative (2008), *Visions for the Canadian Arctic Research Initiative: Assessing the Opportunities* (Ottawa: Council of Canadian Academies, 2008).

Inuit Heritage Trust (2002), 'Report on the Place Names Workshop February 14–15, 2002', Inuit Heritage Trust [website] <http://www.ihti.ca/english/projects.html>, accessed 16 December 2008.

Jacobsson, M. (2004), 'Acquisition of Territory at the Time of Otto Nordenskjöld', in Elzinga et al. (eds).

Kinnvika (2008), 'Change and Variability of Arctic Systems Nordaustlandet, Svalbard–"Kinnvika"' [website of officially endorsed IPY 2007–2009 research programme no. 58] <http://www.kinnvika.net/>, accessed 10 August 2008.

Korsmo, F. and Graham, A. (2002), 'Research in the North American North: Action and Reaction', *Arctic* 55, 319–28.

Krupnik, I. et al. (2005), 'Social Sciences and Humanities in the International Polar Year 2007–2008: An Integrating Mission', *Arctic* 58, 91–101.

Levere, T.H. (1988), 'Vilhjalmur Stefansson, the Continental Shelf, and a New Arctic Continent', *British Journal for the History of Science* 21, 233–47.

—— (1993), *Science and the Canadian Arctic: A Century of Exploration 1818–1918* (Cambridge: Cambridge University Press).

Lüdecke, C. (1993), 'Aspekte zur Institutionalisierung der deutschen Polarforschung von der Jahrhundertwende bis zum II. Weltkrieg', *Zeitschrift für geologische Wissenschaften* 21, 633–40.

McCook, S. (1996), '"It May Be Truth, But It Is Not Evidence": Paul du Chaillu and the Legitimation of Evidence in the Field Sciences', *Osiris,* 2nd series, 11, 177–97.

Merryman, J.H. (1986), 'Two Ways of Thinking About Cultural Property', *The American Journal of International Law* 80, 831–53.

Neumayer, G. von (1901), *Auf zum Südpol! 45 Jahre Wirkens zur Förderung der Erforschung der Südpolar-Region 1855–1900* (Berlin: Vita).

Nilsson, I. (1978), *Grönlandsfrågan 1929–1933: En studie i småstatsimperialism,* Acta Universitatis Umensis: Umeå Studies in the Humanities, no. 17 (Umeå: Umeå University).

Pohjola, V. (ed.) (2006), 'Kinnvika – Arctic Warming and Impact Research', [International Polar Year Programme 2007–09] <http://www.ipy.org/index.php?/ipy/detail/kinnvika>, accessed 10 August 2008.

Powell, R.C. (2002), 'The Sirens' Voices? Field Practices and Dialogue in Geography', *Area* 34, 261–72.

Ries, C. (2002), 'Lauge Koch and the Mapping of North East Greenland: Tradition and Modernity in Danish Arctic Research, 1920–1940', in Bravo and Sörlin (eds).

Rozwadowski, H.M. (1996), 'Small World: Forging a Scientific Maritime Culture for Oceanography', *ISIS* 87, 409–29.

Stefansson, V. (1921), *The Friendly Arctic: The Story of Five Years in the Polar Regions* (New York: Macmillan).

—— (1922), *The Northward Course of Empire* (New York: Macmillan).

Swedish Polar Year Expedition Sveagruvan, Spitzbergen 1932–33: General Introduction, Terrestrial Magnetism (1939), (Stockholm: Swedish National Committee for Geodesy and Geophysics).

Tjelmeland, H. and Zachariassen, K. (eds) (2004), *Inn i riket: Svalbard, Nord-Norge og Norge: Rapport fra det 28. nordnorske historieseminar, Longyearbyen 3.–5. 10.2003*, Speculum Boreale no. 5 (Tromsø: University of Tromsø).

US Board of Geographical Names (2006), 'Antarctic Names', US Geological Survey [website] (updated 14 March 2006) <http://geonames.usgs.gov/antarctic/>, accessed 24 August 2008.

Weiss, K. and Gaud, R. (2006), 'Formation and Transformation of Relational Networks During an Antarctic Winter-Over', *Journal of Applied Social Psychology* 34, 1563–86.

—— et al. (2007), 'Uses of Places and Setting Preferences in a French Antarctic Station', *Environment and Behavior* 39, 147–64.

West, P. (2008), 'A Special Report: U.S. South Pole Station: Navy Station 1956', National Science Foundation [website], (updated 17 January 2008) <http://www.nsf.gov/news/special_reports/livingsouthpole/station56.jsp>, accessed 18 August 2008.

Wijkander, A. (1880), '1881 års föreslagna svenska Spetsbergsexpedition', *Ny svensk Tidskrift* 2, 95–108.

Wråkberg, U. (ed.) (1999a), *Arktisk gruvdrift: Teknik, vetenskap och historia i norr*, Jernkontorets Bergshistoriska utskott, series H no. 69 (Stockholm: Swedish Steel Producers Association).

—— (1999b), 'Politik och vetenskap i A. E. Nordenskiölds ockupationsförsök av Spetsbergen år 1871–1873', in Wråkberg (1999a).

—— (ed.) (1999c), *The Centennial of S. A. Andrée's North Pole Expedition: Proceedings of a Conference on S. A. Andrée and the Agenda for Social Science Research of the Polar Regions*, Bidrag till Kungl. Svenska Vetenskapsakademiens Historia no. 28 (Stockholm: Royal Swedish Academy of Sciences).

—— (2001), 'Eugenio Parent and Giacomo Bove: Italians in 19th Century Swedish Polar Exploration', in Beretta and Grandin (eds).

—— (2002), 'The Politics of Naming: Contested Observation and the Shaping of Geographical Knowledge', in Bravo and Sörlin (eds).

—— (2003), 'Ruiner i förfallande landskap: Arktiska spår av kultur, vetenskap och industri' in Avango and Lundström (eds).

—— (2004a), 'Delineating a Continent of Snow and Ice: Cartographic Claims of Knowledge and Territory in Antarctica in the 19th and Early 20th Century', in Elzinga et al. (eds).

—— (2004b), 'Det internationella intresset för Arktis 1800–1925', in Tjelmeland and Zachariassen (eds).

PART II
Whose Environment? Science and Politics in Antarctica

Chapter 5

Antarctic Science, Politics and IPY Legacies

Julia Jabour and Marcus Haward

Introduction

> Even though science may not be the only factor in management policy, it is at least the necessary basis for such a policy. (Fløistad 1990, 22)

Scientific research is becoming increasingly fundamental in informing political decisions on the effective management of the earth's environment and its resources. Numerous examples illustrate the difficulties of trying to merge science and policy, primarily because of cultural and intellectual differences, first, and evocatively, discussed by C. P. Snow as the 'two cultures' (Snow 1998 [1959]) – leading to the so-called science–policy gap (May 2002). Examples of the science–policy gap include the collapse of the cod fishery off Newfoundland Canada in the 1990s (Finlayson 1994) and more recent debates over climate change. While scientific knowledge and information are used to inform public policy and decision-making (May 2002), scientific research can exacerbate policy disagreements as often as it can resolve them (Hildreth 1994). Consequently, opportunities exist for the gap to be misused in the adoption of decisions which, though favourable politically, may ignore or disregard scientific evidence. The polar management regimes provide useful case studies because while science is a key driver of decision-making, political and diplomatic pressures, including sovereignty and national interests, heavily influence it. This chapter focuses primarily on the Scientific Committee on Antarctic Research (SCAR) and its varying fortunes connected with the Antarctic Treaty System (ATS,[1] Figure 5.1). This is the system of laws founded on the Antarctic Treaty of 1959 and the subsequent instruments regulating sealing, marine living resources conservation and harvesting and environmental matters, that have been adopted by the parties. SCAR is a useful case study for Antarctic science, politics and IPY legacies because of its original role in 1959 as the primary source of independent scientific advice to the Treaty parties; the gradual

---

1 The Antarctic Treaty System (ATS) includes the Antarctic Treaty 1959; the Convention for the Conservation of Antarctic Seals (1970; CCAS); the Convention on the Conservation of Antarctic Marine Living Resources (1982; CCAMLR) and the Protocol on Environmental Protection to the Antarctic Treaty (1991; Madrid Protocol). The text of all documents is available through the Antarctic Treaty Secretariat website <http://www.ats.aq> and its links.

and sometimes subtle redefinition of its powers and functions resulting from the continuing evolution of the ATS; and the opportunity presented to it through the IPY to regain some of this lost ground.

The polar regions (which jointly comprise over 50 million km²) are the antithesis of each other, in more ways than just geographically. The Arctic supports indigenous and non-indigenous communities, large-scale and varied resource exploitation, increasing tensions over sovereignty, relative proximity, unrestricted uses including those of a military nature, major shipping and air transport routes and substantial terrestrial faunal and floral populations. The Antarctic has none of these, except for the controlled exploitation of fish stocks in the Southern Ocean and a small-scale but expanding tourism industry. Paradoxically, the Antarctic has a well-established, coherent regional management regime with legal personality, whereas the Arctic, with its population, resource exploitation and prominent strategic profile, does not have the same cohesive legal regime. The difference largely comes down to one of sovereignty.

Notwithstanding, the primary similarity between the Arctic and the Antarctic is their importance in adding to the scientific understanding of global processes, and of intrinsic polar processes. It is known and widely accepted that human activities are substantially altering global climate patterns by enhancing the earth's natural greenhouse condition (IPCC 2007). Consequently, scientists anticipate a generalized global warming trend and sea level rise, with the greatest physical changes or early indications of such changes being discovered in the polar regions first, although the scope and timing in these conclusions are still highly speculative. Actual scientific information may vary between the Antarctic and the Arctic, due to intrinsic data gathering and interpretation differences and difficulties, but this does not detract from the paramount value of polar science, both discrete and comparative. Together with the oceans and the atmosphere, the poles form an integral part of global climate processes. Therefore, while the

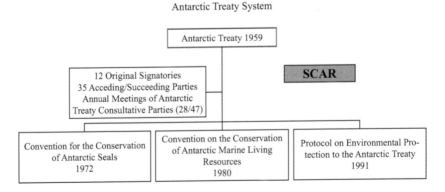

Figure 5.1   The Antarctic Treaty System

Source: Julia Jabour

regions are remarkably opposite in many ways, there are fundamental similarities based on scientific utility and the vulnerability of the polar ecosystems to human interference. Rather than being simply fragile or robust, the polar environments are vulnerable because of:

- low temperatures that retard the breakdown of contaminants;
- short growing seasons, which are a product of cold temperatures, limited sunlight and cold soils, that retard regeneration;
- low biological diversity, a short food chain and high stock levels make species vulnerable to pollution catastrophes;
- highly productive marine areas make the seas and rivers vulnerable to contamination; and
- climatic conditions that favour the deposition or concentration of airborne contaminants in certain areas (Osherenko and Young 1989, 111–16).

Therefore, in a worst-case scenario of enhanced global warming, the major impacts will most likely be felt in the polar regions first. Indeed there is evidence to suggest that Arctic sea ice extent has already been greatly reduced (NOAA 2007) and there is perceptible warming in the Antarctic Peninsula (BAS 2007) and on the surface of the Southern Ocean (SCAR 2007a). Sound management, through a balanced and integrated approach to environmental and developmental questions, is therefore crucial for the polar regions because:

- much of the world's weather is generated in the northern and southern polar regions;
- essential sciences are conducted in these natural laboratories and will lead to a better understanding of global processes;
- the nature of the polar ecosystems is such that they can provide early warning of process changes; and
- there is evidence of a link between climate in the northern and southern polar regions (Steig 2006).

Polar ecosystems are largely still enigmatic to the researcher, however, and little is known about the true extent of interaction between all of the complex components. Polar ecosystems are dynamic (blooms and cycles), diverse and yet specialized (localization of productivity) and at times some species exist at the very limits of survival. Conditions are harsh when compared to temperate climates but many organisms have evolved or adapted to maximize survival. Just as the polar ecosystems are slowly revealing themselves to scientists, climatic changes threaten to significantly alter prevailing conditions and to make scientific research even more important in decision-making.

This chapter examines the contemporary relationship between science and politics in the polar regions, with a particular emphasis on the Antarctic and the role of SCAR and other scientific bodies in informing decision-making. The Antarctic is an exemplar of the science–policy interface because of its highly developed, legally binding regimes, the decisions from which have – theoretically – a foundation in high quality scientific research. To suggest, however, that decisions are based solely on scientific input, or that scientists are always heeded in decision-making, is misleading. Sovereignty and national interest are the defining factors. It is the same in the Arctic, where decisions rely on both scientific input and state will. This chapter illustrates the sometimes uncomfortable coexistence between science and politics, and the machinations of restoring the balance through the IPY.

Antarctic research in earlier polar years

The IGY of 1957–58 is generally seen as giving impetus to the development of the Antarctic Treaty of 1959. At the height of the Cold War the Antarctic Treaty provided a resolution to the Antarctic problem, reconciling differences over the status of claims to territorial sovereignty in Antarctica, with Article IV effectively freezing existing and any further claims during the life of the Treaty (which is unspecified). In addition to demilitarization of the Antarctic continent, the Antarctic Treaty also reinforced the role of science and scientific collaboration, first established through polar discovery, exploration and the coordinated effort of the first and subsequent International Polar or Geophysical Years.

Karl Weyprecht, who was instrumental in the organization of the first IPY of 1882–83, considered that small countries must be able to take part in Arctic research and that results should be freely shared without discrimination because 'science is not a territory for national possession' (Roots 1984, 11). Exploration and scientific research were closely linked to the polar regions, with advancement of knowledge and 'filling out the map' powerful drivers in the eighteenth and nineteenth centuries (Murray 2005).

Weyprecht's idea had been to *coordinate* the steeplechase of polar exploration (as he called it), to *plan* cooperative effort and to *share* observations, even at the expense perhaps of some state prestige (Roots 1984, 11). After much persuasion by Weyprecht and his disciples, an International Polar Commission was established in 1879, principally to coordinate the first IPY. Although Weyprecht died prior to the IPY, its conduct rested on these principles he had first proposed.

Fortuitously, the volcanic island of Krakatau in the Indonesian archipelago erupted after the IPY global observation stations had been established, thereby making it possible to track the atmospheric movement and distribution of Krakatau's volcanic dust clouds over the entire planet. This serendipitous event lent great credence to the concept of cooperative and coordinated scientific effort through events like the IPY and follows a proposal for a world meteorological network by the first International Maritime Meteorological Congress in Brussels in 1853 (Budd 2002, 43). It turned scientific research from an exclusive, domestic pursuit into a commonwealth quest, helping to promote the open and frank peer review that characterizes most scientific disciplines today (Roots 1984, 13).

At the same time, science as an institution was changing. It became more specialized and organized, with groups such as the United Kingdom's Royal Society and Royal Geographical Society having, as their names suggest, support from the highest levels of British society (the monarchy). These, along with similar bodies in other countries (for example, specialized scientific societies in France, Germany and Russia), became important mechanisms in the pursuit of knowledge (Budd 2002, 47). International scientific linkages such as those through the International Meteorological Organization (IMO) were strengthened. The first IPY focused these linkages on the polar areas.

A total of 12 countries conducted a dozen expeditions to the Arctic during the first IPY but only three expeditions went to Antarctica. The second IPY was convened 50 years later in 1932–33 and involved 44 countries. However, the Second World War neutralized many of the achievements of this occasion as records were lost or never written up. Scientific developments during the Second World War (particularly in the areas of rocketry, radar and radio), together with increased awareness of the importance of polar areas for understanding the earth's magnetic field, helped drive a renewed interest in large-scale scientific experiments. This imperative was further enhanced by the successes of the United States' Operations *Highjump* (1946–47) and *Deep Freeze* (1955–56). These expeditions reinvigorated Antarctic science but also inflamed debates over Antarctica's geopolitics (Moore 2004). It was therefore decided to hold another polar year (the IGY) beginning in 1957 (Walton 1987, 32). In the interim, the *Discovery* voyages 1925–39 (*Discovery II* from 1929) and 1950–51 were also undertaken and represented perhaps the most comprehensive marine science investigations of the time.

The International Council of Scientific Unions (ICSU, now the International Council for Science) was in favour of the proposed third polar year and established the Comité Spéciale de l'Année Géophysique Internationale (CSAGI) to coordinate programming and participation. The Antarctic, by virtue of the paucity of existing scientific information, was singled out for special treatment, as was outer space. In 1955 the Soviet Union registered its intention to participate in the Antarctic programs of the IGY and then, in 1957, launched its first *Sputnik* spacecraft (Quigg 1983, 47). These activities by the Soviets caused significant concern throughout the scientific community (Gan 2009), however CSAGI deftly managed to *depoliticize* the science in this instance, despite the undercurrents of disputes between the United States and the Soviet Union (the Cold War), and between Britain, Chile and Argentina (overlapping claims to Antarctic territory), not to mention the state of global politics in general (Hall 1994; Gan 2009).

The ensuing 18 months of research and data collection from July 1957 to December 1958 by 12,000 scientists from 67 countries generated a total of 48 volumes and a collection of scientific papers, the number of which is unknown (Walton 1987, 34; Quigg 1983, 47). This was a considerable achievement given the placement of the IGY within the Cold War power struggle and is testament to the conflict resolution and confidence-building skills of the national scientific entrepreneurs. One commentator, in fact, has suggested that the success of the

IGY might in part be due to the fact that scientists, not governments, were the negotiators and set the programmes and sites (Quigg 1983, 48).

Prior to the IGY, the governments with an interest in the Antarctic had tried and failed to come to agreement on the form of administration for this valuable polar laboratory, though not its need. Thus, in many respects the success of the IGY stimulated the diplomats to begin negotiating in earnest a solution to the problem of Antarctic administration. The fact that the participating governments accepted CSAGI's recommendations on the placement of their scientific stations exemplified their willingness to cooperate during the IGY. The United States, for example, had bases located by CSAGI at the South Pole, and the Soviets, who had originally wanted that site, deferred to CSAGI and established bases elsewhere (Quigg 1983, 48; Gan 2009). Many governments benefited from the international cooperation engendered by the IGY and were keen to see the established programs continue. While it was originally envisaged that the IGY infrastructure and programs would be downgraded or dismantled altogether, both the United States and the USSR announced they were considering retaining a presence on the continent, thereby encouraging each other to commit to long term scientific programs (Hall 1994; Gan 2009). The scene was set for the successful negotiation and adoption of the Antarctic Treaty of 1959.

SCAR – The Antarctic IGY legacy

During the IGY, the Special (later, *Scientifi*) Committee on Antarctic Research, SCAR, was created as a non-governmental coordinating body for the international scientific activity. SCAR was not a political organization. Fifty years on, SCAR again has that role during the fourth IPY. SCAR considers itself 'the primary source of independent scientific advice' to the ATS (SCAR 2007b) although, as will be discussed later, its supremacy in this role has been questioned in recent times, including by its own members.

SCAR is an inter-disciplinary committee of the ICSU and initiates, develops and coordinates scientific research in and about the Antarctic, on global processes as well as astronomy (SCAR 2008). The SCAR members, by agreement, have established a number of standing scientific groups, representing the scientific disciplines active in Antarctic research.[2] SCAR is an observer at Antarctic Treaty Consultative Meetings (ATCMs) and is often invited to provide scientific advice to the parties to help inform their decision-making. A recent example is

---

2   These are the Standing Scientific Group on GeoSciences, Standing Scientific Group on Life Sciences, Standing Scientific Group on Physical Sciences, Joint Committee on Antarctic Data Management, Standing Committee on the Antarctic Treaty System and the Standing Committee on Antarctic Geographic Information. There is also a Standing Committee on Finance. For more information see SCAR website at <hppt://www.scar.org/about> (accessed 3 July 2008).

the recommendation from SCAR that the *Arctocephalus* species (fur seals) be removed from specially protected species listing in the Protocol on Environmental Protection to the Antarctic Treaty (Madrid Protocol) (Jabour 2008). It is also asked to advise the Committee for Environmental Protection (CEP; the institution overseeing compliance with the Protocol on Environmental Protection to the Antarctic Treaty); the Commission for the Convention on the Conservation of Antarctic Marine Living Resources (CCAMLR); and the Advisory Committee on Albatrosses and Petrels (established by the Agreement for the Conservation of Albatrosses and Petrels, ACAP). SCAR has also provided keynote addresses to the ATCM on matters of current interest and concern. At ATCM XXX in 2007, for example, the SCAR lecture was on climate change.[3] At these forums, SCAR, as an official observer, is entitled to submit information and working papers to the parties, either in fulfilment of prior requests for advice or on an *ad hoc* basis.[4]

SCAR has 34 full member countries, along with 8 ICSU scientific union members, and other Associates and Honoraries. Full membership is gained through maintaining an active scientific research programme in Antarctica (SCAR 2008).

Although SCAR has survived into the twenty-first century, it has been criticized for not keeping pace with the growth in its membership, the scale of demands for its scientific expertise or the quality of that advice (Ad Hoc Group 2000, 1). For example, SCAR was asked to conduct an audit of the quality of scientific research being conducted in a high-density location in the Antarctic Peninsula, in response to comments from the report of a joint inspection carried out by the United Kingdom, Australia and Peru (UK 2005). The request raised some quite considerable discussion, especially in relation to issues of sovereignty and independence of scientific endeavour. It was apparent that SCAR was by no means universally accepted as providing the last word on Antarctic science, and SCAR's response is interesting:

> SCAR noted with interest the recommendation in the Inspection Report as well as the various comments by the Parties, that it should carry out an in-situ audit. The recommendation, and the various views expressed, will be considered carefully by SCAR. However, SCAR noted that such an audit would be virtually impossible to carry out given the large number of stations in Antarctica and their widely dispersed location (ATCM XXVIII 2005: para 195).

Even prior to this time, SCAR considered itself almost in competition with the other advisory institutions within the ATS, including the CCAMLR Scientific

---

3 The text of this presentation is available from the Antarctic Treaty Secretariat website at <http://www.ats.aq> (accessed 18 March 2008).

4 An information paper is usually one submitted for information only, whereas a working paper is one submitted by a Consultative Party or requested by them from an invited expert such as SCAR, and usually contains substantive information. Only working papers are translated into all four official languages: English, French, Spanish and Russian.

Committee, the Madrid Protocol's CEP and the Council of Managers of National Antarctic Programs (COMNAP) (Ad Hoc Group 2000, 5). Over a period of time, most probably commencing with the adoption of the Madrid Protocol in 1991 but finally articulated at its 1998 meeting, SCAR increasingly recognized that its mandate as *the* 'independent source of objective scientific information and advice' might be under threat (Thiede 2005, 2). Following an intense period of self-reflection, an Ad Hoc Group on SCAR Organization and Strategy was convened in 1999 to evaluate if SCAR had a role in the twenty-first century (Ad Hoc Group 2000, 5). This was not a difficult task; the Group unequivocally and quickly came to the realization that 'SCAR's mission is still needed, and an organization like SCAR is more important and necessary than ever' (Ad Hoc Group 2000, 5). But systemic inertia had meant that fundamental philosophical and organizational changes were necessary to modernize it, while preserving the best of its traditions. SCAR's history is grounded in the physical sciences (Budd 2002) and this is to some extent made obvious by the political posturing of the various disciplines within its ranks. In fact, it seems that many applied life scientists avoid participation in SCAR, which contrasts with the number of academics who use it, particularly those from the physical sciences disciplines.

The Ad Hoc Group reported in 2000 with a series of eight specific strategies for revitalization and renewal:

- Revitalise SCAR by making SCAR more proactive and update its mission in four areas.
- Engage SCAR delegates, alternates, and officers more actively in SCAR to accomplish SCAR's mission as the preeminent authority on science in Antarctica and surrounding oceans.
- Create more flexible and responsive mechanisms at the operating level to coordinate and plan science and provide scientific advice.
- Improve the planning and decision-making functions in the biennial SCAR cycle.
- Improve SCAR's internal and external communication systems.
- Modernise SCAR's secretariat.
- Engage national Antarctic committees and other adhering bodies to renew their commitment to SCAR.
- Implement change rapidly to maintain and enhance SCAR's position as the authoritative leader for science in Antarctica. (Ad Hoc Group 2000, 7)

SCAR's self-initiated meta-reflection was timely: a fourth IPY planned for 2007–08 would be a natural springboard from which a new-look SCAR could regain its status as the initiator, developer and coordinator of essential twenty-first century polar science.

Science and politics in the Antarctic Treaty System

> Although state-of-the-art scientific knowledge is an important input to decision-making, [their] own advisory groups are inevitably influenced by the fact they are political organisations. (Ad Hoc Group 2000, 8)

The participants in SCAR's five major scientific research programmes (SCAR 2008) are only too well aware of the potential influence of politics on the results of their scientific research, and vice versa. Insofar as SCAR itself as an organization is politicized, however, it is more likely to be only the natural politics of science (e.g. traditional disciplinary rivalry) rather than the politics of diplomacy.

Science is often described as the *currency of influenc* (Herr and Hall 1989) within the ATS because it is seen as activity designed to bolster sovereign claims to territory and it is formally used as the benchmark for obtaining decision-making status. The ATS is centred on the Antarctic Treaty and its subsidiary and complementary instruments. Each instrument is founded on principles of freedom of scientific investigation in, and the sharing of knowledge about, the Antarctic and the Southern Ocean but there is a political component to the decision-making that sometimes overrides scientific considerations.

Using and adapting Oran Young's typology of leadership in international regimes, different leadership approaches and strategies that contribute to the enhancement of influence within the ATS can be identified. *Entrepreneurial leadership* 'uses negotiating skills to influence the manner in which issues are presented ... and to fashion mutually acceptable deals'. *Intellectual leadership* 'relies on the power of ideas to shape ... understand[ing of] the issues at stake'. *Structural leadership* differs from the other two as it is derived from structural power – 'the possession of material resources' (Young 1991, 288). Reputations for excellence in science or commitments to the system are distinct assets. Presence 'at the table', whether at Antarctic Treaty consultative meetings or in scientific working groups, provides further opportunities to influence outcomes, particularly where decision-making is based on consensus. The skills and abilities of individuals may also be critical. This is shown with the emergence of *entrepreneurial* leaders, able to construct consensus and move processes forward (Haward 2001). The decision-making processes of each instrument rely on the input of scientific information, although the final outcome is as likely to be influenced by political motives as it is by the science.

The ATS instruments, and the organizational infrastructure established through them, are also SCAR's playing field. In some it has a prominent role, while in others it has none at all. If SCAR is to reclaim its supremacy as the independent scientific adviser to the ATS, it must locate itself within the niches available to it within the system, as described below. SCAR was established during the IGY and Cold War periods to coordinate international scientific effort; it was not a political body (although there is no doubt that traditional politics between rival disciplines was at play). However, the steadying hand of this seemingly homogenous scientific group gave optimism to the countries with an interest in Antarctica during and after the IGY, and this helped them along the path towards resolution of sovereignty issues – thus also facilitating the continuation of scientific effort in Antarctica. However, it is likely that the somewhat romantic notion of giving scientific research pre-eminence over politics has been overtaken by the political realities of the Antarctic regime. These include claimants wishing to protect potential benefits from claimed territory at some point in the future, with claimant parties using scientific research as a shield of legitimacy. They also include the parties themselves contributing to the global problem of illegal fishing, which challenges the legitimacy and undermines the credibility of CCAMLR – a regime in which SCAR virtually has no role today. Another political reality is that increasingly SCAR became unable to carry out its mandate for both economic and logistical reasons (Final Report ATCM XXVIII, paras 181(b) and 184) and quite possibly for reasons of intellectual capacity as well. This relates directly to the political power the consultative parties have within the system, which will be discussed later. The following section begins by describing the genesis of the Antarctic Treaty and the roles originally and consequently designed for SCAR.

SCAR and the Antarctic Treaty

The development of the Antarctic Treaty was a significant diplomatic effort balancing the aspirations and interests of a number of different actors. Formal negotiations lasted 18 months from the IGY of 1958 and included 60 preparatory meetings. A formal diplomatic conference on Antarctica began in Washington on 15 October 1959 and concluded with the adoption of the Antarctic Treaty on 1 December 1959. In the Preamble it acknowledges 'the substantial contributions to scientific knowledge resulting from international cooperation in Antarctica' and is expresses that 'the establishment of a firm foundation for the continuation and development of such cooperation on the basis of freedom of scientific investigation in Antarctica as applied during the International Geophysical Year accords with the interests of science and the progress of all mankind' (see also Rothwell; Chaturvedi in this volume).

The first ATCM was held in Canberra, Australia from 10–14 July 1961. Addressing the first ATCM, Australian Prime Minister Robert Menzies noted the intense negotiations that had accompanied the drafting of the Antarctic Treaty and commented on the way in which the parties had accommodated different views and interests within this instrument for the pursuit of peaceful scientific exploration. The negotiations of the Treaty took place in, and needed to accommodate, various interests and within the context of the Cold War. The accommodation of the differences between the West and the Soviet Union was, therefore, a great achievement (Hall 1994; Gan 2009). Indeed this had been possible largely because of the effect of the Treaty's Article IV, which freezes existing territorial claims, allowing parties to recognize or not recognize claims as their prerogative, but preventing alterations to the *status quo ante* during the life of the Treaty (the length of which is unspecified). Sovereignty therefore remains a significant, if understated, interest.

The Antarctic Treaty embodies a number of key principles regarding the uses of Antarctica, with Articles II and III providing for the freedom of scientific investigation, the promotion of international cooperation and the free exchange of information and personnel:

Article II
Freedom of scientific investigation in Antarctica and cooperation toward that end, as applied during the International Geophysical Year, shall continue, subject to the provisions of the present Treaty.

Article III
1. In order to promote international cooperation in scientific investigation in Antarctica, as provided for in Article II of the present Treaty, the Contracting Parties agree that, to the greatest extent feasible and practicable:
     (a) information regarding plans for scientific programs in Antarctica shall be exchanged to permit maximum economy and efficiency of operations;
     (b) scientific personnel shall be exchanged in Antarctica between expeditions and stations;
     (c) scientific observations and results from Antarctica shall be exchanged and made freely available.
2. In implementing this Article, every encouragement shall be given to the establishment of cooperative working relations with those Specialized Agencies of the United Nations and other international organizations having a scientific or technical interest in Antarctica.

SCAR was not mentioned explicitly, but there is no doubt that it is one of those *other international organizations* with an interest in Antarctic science.

The 12 original signatories to the Treaty[5] – the states that were most active in Antarctica during the IGY – have decision-making status at ATCMs by virtue of Article IX of the Treaty. The original signatories do have some shared interests (primarily in maintaining their elite position with regard to claims or rights to claim), but also some competing and conflicting interests. Article XIII provides for accession to the Treaty by other states and they too can achieve consultative party status by 'conducting substantial research activity [in Antarctica], such as the establishment of a scientific research station or the despatch of a scientific

---

5   The key stakeholders in Antarctica are the claimant states: Argentina, Australia, Chile, France, New Zealand, Norway and the United Kingdom; the United States and the Russian Federation (succeeding the Soviet Union), both of whom reserve their rights to make claims; and the other original signatories, none of whom have made claims: Belgium, Japan and South Africa.

expedition' (Article IX.2). Since the adoption of the Protocol on Environmental Protection to the Antarctic Treaty in 1991 (the Madrid Protocol), these prerequisites have been broadened to allow interested parties to benefit from the support of other national programs without being required to add to the built infrastructure on the continent. Of the current 47 parties to the Treaty, 28 have decision-making status (ATS 2008).

The fact that the Treaty did not create any institutions per se, apart from the ATCM itself, has been problematic for the support of Antarctic affairs, including scientific research. For example, it was not until 2003 that consensus was reached[6] on the establishment of a Secretariat, the need for which was resisted for many years. That argument was no longer defensible after the adoption of the Madrid Protocol and the inherently bureaucratic workload of the CEP (see below). Thereafter the location of the Secretariat became a political football and further delayed progress until consensus could be reached. The Antarctic Treaty Secretariat has been located in Buenos Aires since 2004 and has responsibilities for supporting the ATCMs and the CEP; promoting official information exchanges between the parties; collecting, maintaining and publishing the records of the ATCM and the CEP; and providing information on the Treaty system (ATS 2008). The latter is the first official educational outreach program of the ATCM, which is traditionally hosted by a consultative party and rotated in alphabetical English order – a process which in the past was an exclusive, expensive and variously successful approach facilitating neither useful information exchange nor a 'shop front' for Antarctic business or science.

Antarctic science, particularly that carried out under the auspices of the IPY (see following) is characterized by international collaboration in both research and logistics. International programs in glaciology, ice coring, upper atmospheric physics, astronomy and marine science are examples here. Many scientific programs rely on collaboration, and scientific exchanges with other countries are a regular feature of national Antarctic programs. Other activities in the Antarctic show the links between interests of claimant states and the broader commitment to scientific endeavour. Obtaining the coordinates for delimiting an extended continental shelf zone off the Australian Antarctic Territory, for example, utilized a range of provisions available to Australia under Article 76 of the United Nations Convention on the Law of the Sea (UNCLOS), including the sediment thickness formula. It is important to note that Australia asked the Commission on the Limits of the Continental Shelf not to examine this data in its 2004 submission (Australia 2004; Jabour 2006), despite the large sum of money expended on collecting it. This was explicit recognition of the sensitivities of a narrow interpretation of Article IV of the Antarctic Treaty with respect to enlargement of existing claims. Australia's concession to the spirit of harmony within the Antarctic Treaty was a diplomatic masterstroke; notwithstanding, its efforts also provided significant

---

6   The Antarctic Treaty consultative parties make their decisions by consensus, which is taken to mean either agreement or the lack of formal objection.

scientific data for future research on climate change and potential resources of the seabed and subsoil.

The Antarctic Treaty is elegantly simple, providing an interim solution to the problem of Antarctic sovereignty and a range of highly acceptable philosophical positions on the egalitarian value of peace and scientific endeavour in the region. While SCAR was not mandated expressly with the role of scientific coordinator and adviser within the Articles of the Treaty, it no doubt inherited a significant role as a legacy of its IGY responsibilities. As the Antarctic legal regime developed, SCAR was provided with opportunities to expand its mandate in Antarctic science.

SCAR and the Convention for the Conservation of Antarctic Seals (CCAS)

CCAS not only aims to protect seals and maintain ecosystem balance, but also has the typical living resources objective of managing commercial harvesting at sustainable levels by prescribing certain measures based on the reporting and free exchange of scientific research. Its intention in this regard is clear:

> Recognizing that in order to improve scientific knowledge and so place exploitation on a rational basis, every effort should be made both to encourage biological and other research on Antarctic seal populations and to gain information from such research and from the statistics of future sealing operations, so that further suitable regulations may be formulated. (Preamble)

The area of application of CCAS is the sea south of 60° South (Article 1) – a political boundary consistent with the area of application of the Antarctic Treaty. There is a domestic permit system for the taking of seals for food (removed from the later 1991 Madrid Protocol), research and specimen collection, and parties are accountable for any take by reporting to the other parties and to SCAR (Article 4). The CCAS does not establish an independent scientific advisory body but relies exclusively on the participation of SCAR. Article 5 requires that the parties exchange information and scientific advice between them, as provided for in Section 6 of the Annex. Here it asks parties to provide a range of information including, for example, 30 days' advance notice of the departure from home ports of sealing expeditions, which would signal the re-commencement of commercial harvesting. SCAR is given the responsibility of overseeing the scientific aspects of harvesting and conservation, including advising on the need to cease harvesting if necessary. In the event that commercial sealing resumes, the parties would meet to establish the rules for monitoring, control and surveillance of activities carried out under national and flag state jurisdiction (Article 6).

Because there is no commercial harvesting of seals in the Southern Ocean, and no likely prospect of resumption in the near future, CCAS is inert at the moment. The decision to de-list all *Arctocephalus* species from the Madrid Protocol's Annex of specially protected species, which was a SCAR recommendation, is probably the only significant issue on the CCAS radar. The parties may, at some point, decide to convene a meeting to discuss removing these species from their CCAS special protection as well, for the sake of consistency (Jabour 2008).

SCAR and the Convention on the Conservation of Antarctic Marine Living Resources (CCAMLR)

Until the 1980s, SCAR had exclusive competence over Antarctic scientific matters through its IGY inheritance and its explicit role in CCAS. However with the adoption of CCAMLR in 1980, the institutions created by it diluted SCAR's role somewhat. Notwithstanding, scientific advice is central to CCAMLR. The Convention applies to Antarctic marine living resources (excluding whales and seals) in an area bounded to the north by a line that approximates the Antarctic Convergence (now termed the Antarctic Front) and to the south by the Antarctic continent (Article I) – a biogeographical rather than political boundary. A commission (Article VII) takes advice from its scientific committee (Article XIV) and applies a precautionary and ecosystem-wide approach in formulating management provisions for the conservation and rational use of Antarctic marine living resources. CCAMLR currently has 34 states parties (CCAMLR 2008).

The commission's formal function is to facilitate research, compile data, acquire statistics, analyze, disseminate and publish information, identify conservation needs, analyze effectiveness of conservation measures and formulate, adopt and revise conservation measures (Article IX). It is also responsible for implementing observation and inspection systems. Conservation measures may contain any directives that will assist the commission to give effect to the objective and principles of the Convention's Article II.

SCAR has no *formal* role in CCAMLR other than as an observer at its annual commission meetings. Rather, a scientific committee is established as a consultative body to the commission and increasingly the commission relies on its input, rather than that of SCAR, in its decision-making. All members of the commission (currently 25 of the 34 parties) are also members of the scientific committee. The role of the latter is to study the Southern Ocean ecosystems and make recommendations to the commission on appropriate conservation measures. The scientific committee has various standing and ad hoc working groups to assist in its research, analysis and recommendations. Information from national scientific programs is fed into the working group process, and then through each individual group report into the scientific committee. From there an integrated report is compiled for the commission, which finally determines (by consensus) the substance of conservation measures.

Because the activities regulated by CCAMLR are both scientific and economic in nature, that is, research to establish quotas for the commercial harvest of fish, there is significant potential for national agendas to enter into and influence the decision-making process. The high value of the Patagonian toothfish fishery in the

Southern Ocean within the CCAMLR area, for example, has attracted considerable interest and a concomitant rise in what has been termed illegal, unreported and unregulated (IUU) fishing. The level and extent of IUU fishing has posed significant challenges for the commission which has addressed these problems with a number of conservation measures, including the introduction of a catch documentation scheme that tracks the catch and landing of toothfish.

Substantial scientific research activity has identified classic characteristics of the Patagonian toothfish that make it vulnerable to overfishing: slow to mature, long-lived, aggregating in nature, and so on. Obvious strategies adopted by CCAMLR to protect the fish, and the commercial viability of the fishery, include using a vessel monitoring system (VMS) to locate and track licensed fishing vessels. This works on the assumption that every vessel that is identified by VMS and other satellite surveillance techniques, or by first-hand visual sighting, can be placed as either licensed or IUU. CCAMLR scientific working groups collect data and conduct research in order to recommend sustainable regional and smaller scale quotas and for these to be realistic, they must be discounted by a factor representing an estimated IUU catch. It is therefore in everyone's best interests to heed the scientific advice and the recommended quotas. The merit of that advice is for the most part unchallenged until political issues take supremacy, which they very often do as IUU is a political issue with both social and environmental consequences. The attempt by CCAMLR commission members to place vessels flagged to other members on a blacklist for alleged IUU activities is a case in point (Turner et al. 2008).

SCAR and the Protocol on Environmental Protection to the Antarctic Treaty (Madrid Protocol)

Science is similarly embedded within the Madrid Protocol. Article 2 of the Protocol states:

> The Parties commit themselves to the comprehensive protection of the Antarctic environment and dependent and associated ecosystems and hereby designate Antarctica as a natural reserve, devoted to peace and science.

The Madrid Protocol was negotiated following Australia's and France's refusals to sign the Convention on the Regulation of Antarctic Mineral Resource Activities (CRAMRA) in 1989 on the spurious grounds that its built-in environmental protection regime was inadequate. Two years later, a protocol to the Antarctic Treaty was adopted (4 October 1991), with much of the environmental language of CRAMRA transferred directly to the new instrument. Its text and four annexes (Environmental Impact Assessment, Conservation of Antarctic Fauna and Flora, Waste Disposal and Waste Management, and Prevention of Marine Pollution) entered into force on 14 January 1998. A fifth annex (Area Protection and

Management) that had a separate approval process did not enter into force until 24 May 2002. The Protocol included a commitment to develop rules relating to liability for environmental damage (Article 16). There were protracted negotiations over liability that began during the CRAMRA negotiations and concluded nearly 20 years later, with ATCM XXVIII in Stockholm in June 2005 finally adopting the Protocol's sixth annex: Liability Arising from Environmental Emergencies. This modest attempt is considered the first stage of what is expected to become a more comprehensive Antarctic liability regime in time. Annex VI is not yet in force.

The focus on environmental protection has encouraged new, multi-disciplinary science programs amongst Treaty parties (Haward et al. 2006). The Protocol aims to limit adverse impact and effect on the total ecosystem through a mandatory environmental impact assessment scheme (Protocol Article 8 and Annex I). Activities (including scientific research projects) planned by ATCPs, and by tourist and non-governmental groups (though not fishers, because of an agreement to give CCAMLR competency over all matters related to fisheries), must take account of the scope of the activity, cumulative impacts, safety, capacity to monitor and capacity to respond promptly to accidents, which must be reported within an environmental evaluation. However, the terms *scope* and *cumulative impact* are not defined, leaving interpretation largely arbitrary according to state practice, but with some cursory guidelines developed by the parties through the CEP.

The CEP was established under Article 11 of the Protocol to advise the ATCM on the Protocol's implementation. Its meetings and intersessional working groups are usually composed of a mixture of environmental officers, scientists and diplomats. The CEP reviews all environmental matters and reports to ATCMs with recommendations. It seeks advice on an ad hoc basis from various specialist groups, including SCAR and other scientific organizations such as the International Hydrographic Office, the World Meteorological Organization (WMO) and the International Union for the Conservation of Nature and Natural Resources (IUCN). The CEP maintains a database of Initial Environmental Evaluations and Comprehensive Environmental Evaluations (CEEs) prepared by the parties. Of these mandatory environmental assessments, the CEP only formally reviews CEEs. After a process of party and CEP scrutiny, the CEP makes recommendations to the ATCM about the status of the evaluation, but the ATCM cannot veto any activity (nor delay it for longer than 15 months). The case of a proposal by India to establish a new scientific research station in the Larsemann Hills is illustrative. The Larsemann Hills are recognized for their unique and sensitive environmental features. India proposed to build a new station outside a designated facilities area (containing other stations and infrastructure) established under a management plan being developed for the Larsemann Hills by Australia. The CEE for the Indian station raised concerns, and was the subject of significant discussion, at the 2006 ATCM. A lack of consensus at this meeting led to the proposal being re-worked and resubmitted to, and approved by, the 2007 ATCM. While there were some modifications to the station's design, the proposed site was maintained.

In summary, the scientific institutions established by instruments of the ATS are essential from a practical perspective, but they are primarily advisory in nature. While their scientific research might figure prominently in the rhetoric of the regime, the Realpolitik is that the power belongs to the diplomats and policy makers. However, threats to polar environments from human and environmental changes, such as those caused by climate change and technological developments, are leading to increasing challenges for governance of polar areas. The need to assess, quantify, understand and communicate these impacts, changes, challenges and threats has been recognized by the announcement of the current IPY. The IPY is occurring at a time of great significance for the big questions on climate change and the question is posed here as to whether it will raise the profile of Antarctic (and indeed polar) science in the same way that its forerunner, the IGY, did. However, Walton notes that 'what is lacking this time is the same overt political enthusiasm and pubic interest that could make this [IPY] a milestone like IGY' (Walton 2007, 1).

Antarctic science and the IPY

The IPY again provides the potential for significant impetus to Antarctic scientific research and its input into the policy process, with a commitment of over $1.25 billion (of which about one-third is new investment) indicating the level of international interest (ATCM XXX 2007, para 125). In a joint paper presented to ATCM XXX in 2007, SCAR and the International Programme Office of the IPY estimated, however, that a further several hundred million dollars would be required to fulfil all commitments undertaken in IPY-endorsed projects, along with the building of new research vessels and new or refurbished Antarctic infrastructure (ATCM XXX 2007, para 125). SCAR is either leading or involved in 70 per cent of the polar or Antarctic natural science, IPY-endorsed projects.

The many endorsed programs on offer (about 228) were initiated by Arctic states and Antarctic Treaty states individually and jointly. In addition, there is a host of global collaborative projects. Importantly in this modern era, nearly a quarter are specific projects involving education and outreach. This unprecedented attention to the social sciences, humanities and communication brings scientific research out of the realm of the laboratory and into the hearts and minds of a wide public audience (including policy makers), which will both benefit from scientific results and be asked to make sacrifices towards global solutions to transboundary environmental problems like climate change.

One of the projects most likely to achieve a high public and scientific profile is the Census of Marine Life. This involves a global network of researchers from more than 70 countries engaged in a long-term initiative to assess and explain the diversity, distribution and abundance of marine life in the oceans (COML 2008). During the IPY they will be conducting research on the Southern Ocean. This kind of science, along with dedicated programs in all scientific disciplines, is vital to a better understanding of physical and biological processes world-wide, including in both polar regions.

The Census of Marine Life is a major ship-based research program that took place during the southern hemisphere summer of 2007–08. Scientists from about 30 countries and 50 institutions on up to 16 research vessels conducted various surveys in Antarctic waters. Some projects were also conducted from onboard tourist vessels. The Census has been the most comprehensive survey of the distribution and abundance of Antartic Marine Life ever conducted and it will provide a benchmark for tracking future change in the Antarctic marine environment (Hosie et al. 2008). The Census of Antarcic Marine Life is a key SCAR activity, which goes back to the basics of developing a 'robust benchmark', and will leave a legacy of 'a fully operational and interoperable database' (Australia/SCAR 2007). Without such benchmark data, it would be almost impossible to have a position of any accuracy from which to begin to measure climate impact on the Southern Ocean ecosystems.

SCAR has a pivotal role in this, and other IPY projects. The question remains as to whether this IPY will do what the other polar years have done in terms of reinvigorating and helping to set directions for the coming decades of scientific research. Improved technology and communications will go a long way towards ensuring that information, results and observations from this IPY are widely and quickly disseminated. The legacy of the IPY is already looking positive. At the 30th ATCM in 2007, the parties passed a resolution (3/2007), which reads thus:

Long-term Scientific Monitoring and Sustained Environmental Observation in Antarctica

The Representatives,

Recalling the Edinburgh Antarctic Declaration on the International Polar Year 2007–2008 (IPY) that was agreed at ATCM XXIX, which supports the objective of delivering a lasting legacy for the International Polar Year, and promotes increasing collaboration and coordination of scientific studies within Antarctica;

Recalling that the Committee for Environmental Protection has a continuing commitment to environmental monitoring related to the implementation of the Protocol on Environmental Protection to the Antarctic Treaty;

Noting that the Arctic Council Ministerial Meeting of 26 October 2005 urged all member countries of the Arctic Council to maintain and extend long-term monitoring of change in all parts of the Arctic as well as to create a coordinated Arctic observing network;

Recalling the success of the CCAMLR Ecosystem Monitoring Programme in providing over two decades of circum-Antarctic data on the Antarctic marine ecosystem and biological environment;

Welcoming and supporting the proposal by the Scientific Committee for Antarctic Research to establish a multi-disciplinary pan-Antarctic observing system, which will, in collaboration with others, coordinate long-term monitoring and sustained observation in the Antarctic;

Recommend that the Parties:

1. urge national Antarctic programmes to maintain and extend long-term scientific monitoring and sustained observations of environmental change in the physical, chemical, geological and biological components of the Antarctic environment;

2. contribute to a coordinated Antarctic observing system network initiated during the IPY in cooperation with SCAR, CCAMLR, WMO, GEO and other appropriate international bodies;

3. support long-term monitoring and sustained observations of the Antarctic environment and the associated data management as a primary legacy of the IPY, to enable the detection, and underpin the understanding and forecasting of the impacts of environmental and climate change.

While some IPY research projects are short term, others run the length of the IPY and beyond. At some point in the future it will be possible to do a comprehensive audit of the IPY, putting numbers of researchers, ships, projects, collaborations, outreach, education and expenditures into a matrix to assess its effectiveness. At a further point, it will be possible to look back and measure the durability of the legacies from this IPY. For Walton 'the most important legacy' would be that the polar regions are seen as a seamless yet critical part of the global picture, that their science will more closely influence international policy and their conservation and good management will be seen as a political priority by all the countries involved. These may be less tangible outcomes than the objectives on the IPY website but they could be of greater long-tem importance (Walton 2007, 1).

Discussion and conclusions

Up to now, science and policy have operated largely as two separate communities, some would say with unique values. But we can no longer work as if in a relay race – a scientist completes a piece of work and then passes it off to a policy person to run the next leg of the race. More like a rugby team, scientists and policy analysts must run the field together, supporting each other as they go, and achieving goals as a united team. (May 2002, 1)

In a report to ATCM XXX in 2007, SCAR was effusive in describing potential legacies from the IPY (SCAR 2007c). On behalf of the organizing sponsors, SCAR reported that from the start the IPY was seen as 'a unique and compelling opportunity to develop sustained observing systems at both poles' (SCAR 2007c). With such a heavy burden of responsibility imposed on the parties by their adoption of the Madrid Protocol, it could be argued that without the baseline data collected by the kind of monitoring referred to by SCAR, and by projects such as CAML, they have virtually no hope of ever meeting the obligations under Article 3 of the Protocol, which reads in part:

2. To this end:
   (b) activities in the Antarctic Treaty area shall be planned and conducted so as to avoid:
      (i) adverse effects on climate or weather patterns;
      (ii) significant adverse effects on air or water quality;
      (iii) significant changes in the atmospheric, terrestrial (including aquatic), glacial or marine environments;
      (iv) detrimental changes in the distribution, abundance or productivity of species of populations of species of fauna and flora;
      (v) further jeopardy to endangered or threatened species or populations of such species; or
      (vi) degradation of, or substantial risk to, areas of biological, scientific, historic, aesthetic or wilderness significance;

   ...

   (d) regular and effective monitoring shall take place to all assessment of the impacts of ongoing activities, including the verification of predicted impacts;
   (e) regular and effective monitoring shall take place to facilitate early detection of the possible unforeseen effects of activities carried on both within and outside the Antarctic Treaty area on the Antarctic environment and dependent and associated ecosystems.

In acknowledgement that the earth is an integrated system, the SCAR report emphasized the need for global–polar observing systems at sustained levels over the long term, and this was supported by the adoption of Recommendation 3 from the 2007 ATCM. In addition, SCAR considered other legacies as being vast steps forward in data and information, and improved skills of global and regional weather and climate models. Good models are better from high-quality observations, and while coupled atmosphere-ocean climate models are seen as the main tools to help predict future climate evolution, they are assessed by SCAR as currently struggling to represent key aspects of polar climate (SCAR 2007a).

SCAR reported there are some aspects of the Antarctic Peninsula climate that can be described reasonably confidently: near-surface warming, rapid ocean surface warming, glacial retreat and ice-shelf collapse. On the other hand, there has been much less recent ice loss from East Antarctica, which has also not shown significant warming. In fact there appears to have been an increase in sea ice in the Ross Sea region. The troposphere has warmed while the stratosphere has cooled and westerly winds have intensified in the Southern Ocean. Climatic projections for Antarctica over the next 90 years are broad: warming of the sea ice zone, a reduction in sea ice extent, a warming of the Antarctic interior and increased interior snowfall. However, all of these characteristics of the Antarctic and Southern Ocean climate system reported by SCAR require significant research into the future, and the establishment of the integrated monitoring system will make a significant contribution towards improving the quality of information translated to policy-makers (SCAR 2007c).

Legacies such as new science funding paradigms, the recruitment of a new generation of polar researchers, increased public and political awareness and possibly even political and economic cooperation through capacity building and outreach programs have all been identified (SCAR 2007c). Furthermore, SCAR perceives itself as the conduit for these legacies, by providing continued leadership and energy of the kind it inherited from the IGY but lost during the transition into the twenty-first century.

In this modern age of collaboration, the scientific output from the IPY will add crucial information to our store of knowledge about global and polar processes. Both the Arctic and the Antarctic, where scientific collaboration is already firmly established as a fundamental principle, will play a significant role. Collaboration in scientific endeavour is no longer negotiable, given the role that human activities worldwide play in substantially altering global climate patterns.

The IGY of 1957–58 is credited with giving impetus to the Antarctic Treaty of 1959, even in times of great political turmoil. In the Arctic, geopolitical agendas dominated the discourse for many years, but here, too, collaboration for the benefit of all is the norm rather than the exception. The IPY 2007–08 provides a significant new momentum for international collaboration and coordination in polar science, which will result in benefit sharing, relationship building, maximum scientific outcomes and cost-effectiveness. Polar processes are transboundary and the scientific challenges are often beyond the capacity of one country alone. Some IPY programs, such as the Global Ocean Observing System, are international. Others, such as the Census of Marine Life described here, are global in scope but will be carried out at a regional level.

As noted above we regard the Antarctic as an exemplar of the science–policy interface. The longstanding legally binding instruments that underpin the ATS have reinforced the importance of high-quality scientific research. This does not mean that all decisions are based solely on scientific input, or that scientists are always heeded in decision-making. Even in a regime that emphasizes collaboration and consensus, where sovereignty has been set aside under Article IV of the Antarctic

Treaty, national interests are the defining factors, most notably reflected in debates within CCAMLR. While the geo-political situation in the Arctic is, to coin a phrase, a polar opposite, decisions too rely on both scientific input and state will. The IPY will reinforce and perhaps reinvigorate the nexus between science and politics in the polar regions.

References

Ad Hoc Group on SCAR Organization and Strategy (2000), 'Scientific Committee on Antarctic Research: Preparing Scar for 21st Century Science in Antarctica', Report to the Scientific Committee on Antarctic Research. Hard copy on file with authors.

ATCM XXVIII (2005), Antarctic Treaty Consultative Meeting XXVIII, Final Report <http://www.ats.aq>, accessed 12 February 2008.

ATCM XXX (2007), Antarctic Treaty Consultative Meeting XXX, Final Report <http://www.ats.aq>, accessed 12 February 2008.

ATS (Antarctic Treaty Secretariat) (2007), ATCM XXIX Final Report, paras 85–86 and Measure 4 (2006) <http://www.ats.aq>, accessed 6 February 2007.

ATS (Antarctic Treaty Secretariat) (2008), Home page <http://www.ats.aq>, accessed 6 February 2007.

Australia (2004), Government of Australia, Executive Summary, Submission to the Commission on the Limits of the Continental Shelf. United Nations Division for Ocean Affairs and the Law of the Sea, <http://www.un.org/Depts/los/clcs_new/submissions_files/submission_aus.htm>, accessed 18 March 2008.

Australia/SCAR (2007), Government of Australia and SCAR, Information Paper 32 to ATCM XXX, 'Census of Antarctic Marine Life (CAML)'.

BAS (British Antarctic Survey) (2007), 'Is Antarctica Melting Because of Global Warming?' <http://www.antarctica.ac.uk/About_Antarctica/FAQs/faq_02.html>, accessed 18 March 2007.

Budd, W. (2002), 'Scientific Imperative for Antarctic Research', in Jabour-Green and Haward (eds).

CCAMLR (Commission for the Conservation of Antarctic Marine Living Resources) (2008), Official Member Contacts <http://www.ccamlr.org/pu/E/ms/contacts.htm>, accessed 17 February 2008.

COML (Census of Marine Life) (2008), Home page <http://www.coml.org>, accessed 2 February 2008.

Finlayson, A.C. (1994), *Fishing for Truth: A Sociological Analysis of Northern Cod Stock Assessment from 1977–1990* (St Johns: Institute of Social and Economic Research, Memorial University of Newfoundland).

Fløistad, B. (1990), 'Communication between Science and Decision Makers: The Advisory Function of the International Council for the Exploration of the Sea', *International Challenges* 10:4, 22.

Gan, I. (2009), 'Will the Russians Abandon Mirny to the Penguins after 1959 ... Or Will They Stay?', *Polar Record* 45:233, 167–75.

Hall, R. (1994), 'International Regime Formation and Leadership: The Origins of the Antarctic Treaty', Unpublished PhD Thesis, University of Tasmania.

Handmer, J. (ed.) (1989), *Antarctica: Policies and Policy Development* (Canberra: Centre for Resource and Environmental Studies, Australian National University).

Haward, M. (2001), 'Assessing Influence Within the Antarctic Treaty System', *Australian Antarctic Magazine* 2:Spring, 44.

——Haward, M. Rothwell, D.R., Jabour, J., Hall, R., Kellow, A., Kriwoken, L., Lugten, G. and Hemmings, A. (2006), 'Australia's Antarctic Agenda', *Australian Journal of International Affairs* 60:3, 439–56.

Herr, R. and Hall, R. (1989), 'Science as Currency and the Currency of Science', in Handmer (ed.).

Hildreth, R. (1994), 'The Role of Science in US Marine Policy: Some Regional Applications', *Coastal Management Journal* 22, 163–70.

Hosie, G., Stoddart, M., Wadley, V., Koubbi, P., Ozouf-Costax, C., Ishimaru, T. and Fukuchi, M. (2008), The Census of Antarctic Marine Life and the Australian-French-Japanese CEAMARC contribution <http://polaris.nipr.ac.jp/~ipy/sympo//proc-files/16-Hosie.pdf>, accessed 2 March 2008.

IPCC (Intergovernmental Panel on Climate Change). (2007), 'IPCC/TEAP Special Report, Safeguarding the ozone layer and the global climate system: Issues related to hydrofluorocarbons and perfluorocarbons, Summary for Policymakers' <http://www.ipcc-wg1.ucar.edu>, accessed 3 June 2007.

Jabour-Green, J. and Haward, M. (eds) (2002), *The Antarctic: Past, Present and Future* Antarctic CRC Research Report #28 (Hobart: Antarctic Cooperative Research Centre).

Jabour, J. (2006), 'High Latitude Diplomacy: Australia's Antarctic Extended Continental Shelf', *Marine Policy* 30:2, 197–98.

—— (2008), 'Successful Conservation – Then What? The De-listing of *Arctocephalus* Fur Seal Species in Antarctica', *Journal of International Wildlife Law and Policy* 11:2, 1–29.

May, A. (2002), *Creating Common Purpose: The Integration of Science and Policy in Canada's Public Service* (Ottawa: Canadian Centre for Management Development).

Moore, J.K. (2004), 'Bungled Publicity: Little America, Big America and the Rationale for Non-Claimancy, 1946–1961', *Polar Record* 40:212, 19–30.

Murray, C. (2005), 'Mapping Terra Incognita', *Polar Record* 41:217, 103–12.

NOAA (National Oceanic and Atmospheric Administration) (2007), 'Arctic Change', <http://www.arctic.noaa.gov/detect/ice-seaice.shtml>, accessed 12 March 2008.

Osherenko, G. and Young, O. (1989), *The Age of the Arctic* (Cambridge: Cambridge University Press).

Quigg, P.W. (1983), *A Pole Apart* (New York: McGraw-Hill).

Roots, E.F. (1984), 'International and Regional Cooperation in Arctic Science: A Changing Situation', Paper presented at the Nordisk Konferanse om Arktisk Forskning, held at Ny Ålesund, Svalbard, 2–8 August 1984.

SCAR (Scientific Committee on Antarctic Research) (2007a), Information Paper 5 to ATCM XXX (2007) 'State of the Antarctic and Southern Ocean Climate System (SASOCS)', Antarctic Treaty Secretariat <http://www.ats.aq>, accessed 14 March 2008.

SCAR (Scientific Committee on Antarctic Research) (2007b), Information Paper 6 to ATCM XXX (2007) 'SCAR Report to ATCM XXX', Antarctic Treaty Secretariat <http://www.ats.aq>, accessed 14 March 2008.

SCAR (Scientific Committee on Antarctic Research) (2007c), Information Paper 73 to ATCM XXX (2007) 'IPY Report for ATCM XXX', Antarctic Treaty Secretariat <http://www.ats.aq>, accessed 14 March 2008.

SCAR (Scientific Committee on Antarctic Research) (2008), Home page <http://www.scar.org>, accessed 10 February 2008.

Snow, C.P. (1998 [1959]), *The Two Cultures* (with an Introduction by Stefan Collini) (Cambridge: Canto – Cambridge University Press).

Steig, E. (2006), 'The South–North Connection', *Nature* 444:9, 152–53.

Thiede, J. (2005), 'Science in Antarctica and the Southern Ocean: The new structure of SCAR, the upcoming SCAR conference 2004, some aspects of the research program of the Alfred Wegener Institute for Polar and Marine Sciences in high southern latitudes, and the perspectives for a new IPY', in Proceedings of the 2nd Malaysian International Seminar on Antarctica (2004) *Antarctica: Global Laboratory for Scientific and International Cooperation* (Akademi Sains Malaysia).

Turner, J., Jabour, J. and Miller, D.M. (2008), 'Consensus or not Consensus: That is the CCAMLR Question', *Ocean Yearbook* 22, 117–58.

United Kingdom (2005), Working Paper 32, 'Report of Joint Inspections under Article VII of the Antarctic Treaty and Article 14 of the Environmental Protocol' in ATCM XXVIII Final Report, available from <http://www.ats.aq>, accessed 29 May 2008.

Walton, D.W.H. (ed.) (1987), *Antarctic Science* (Cambridge: Cambridge University Press).

—— (2007), 'Editorial – International Polar Year', *Antarctic Science* 19:1, 1.

Young, O.R. (1991), 'Political Leadership and Regime Formation: On the Development of Institutions In International Society', *International Organization* 45:2, 281–308.

Chapter 6

The IPY and the Antarctic Treaty System: Reflections 50 Years Later

Donald R. Rothwell

Introduction

The IGY was and remains pivotal to an understanding of the governance and management of the Antarctic continent. Notwithstanding rising tensions in Antarctica with respect to the unresolved status of various claims to territorial sovereignty, let alone the wider global tensions associated with the Cold War, the IGY proved to be the ultimate foundation for the various states with an interest in Antarctic affairs to be able to come together in the name of scientific research and work collaboratively together on the continent. This proved to be the catalyst needed in the ongoing debates over the future of Antarctica and was the spur to the eventual November 1959 gathering together of interested states in Washington to negotiate the Antarctic Treaty.[1] The Treaty, which in itself has provided the foundation for the development of the so-called 'Antarctic Treaty System' (ATS) which now includes a framework of additional conventions and instruments in addition to the decisions and recommendations adopted at Antarctic Treaty Consultative Meetings (ATCMs) (Rothwell and Davis 1997), has over five decades proven itself capable of promoting the freedom of scientific research and thereby continue the spirit of the IGY. The 2007–2008 IPY continues that tradition.

However, whilst the importance of science in Antarctica was most remarkably imbedded into the terms of the Antarctic Treaty,[2] and whilst the freedom and promotion of scientific research has continued throughout the duration of the Treaty, there is the potential that the manner in which scientific research has been conducted in Antarctica over the decades may ultimately prove to be the basis for discord amongst the Antarctic Treaty parties. One particular scientific activity is that of bioprospecting. Antarctic bioprospecting, both on the continent and in the Southern Ocean, has the potential to result in important new discoveries but also significant commercial gain for those corporations funding some of the research (Hemmings and Rogan-Finnemore 2004). While it is arguable that a form of bioprospecting has been taking place in Antarctica ever since scientific

---

1   The Antarctic Treaty, done in Washington, 1 December 1959, in force 23 June 1961, 402 United Nations Treaty Series 71.
2   Antarctic Treaty, Article II.

research expeditions were sent to the region over 200 years ago, the important contemporary distinction is the current commercial nature of these activities, thereby raising the issue as to whether bioprospecting is a legitimate research activity, or if it has such significant commercial consequences that it demands regulation, and if so by whom? Another controversial form of scientific research taking place in the region, and specifically the Southern Ocean is with respect to whales. A prominent global controversy has developed into the manner by which Japan undertakes a 'scientific whaling' program in the Southern Ocean notwithstanding regular protests against this activity by not only the International Whaling Commission (IWC) (Wansborough 2004), the international organization which regulates pelagic whaling, but also from Antarctic Treaty states who object to Japan's activities within their proclaimed waters offshore the continent (Blay and Bubnna-Litlic 2006). At the heart of this debate is the argument that Japan's research program is driven by commercial considerations, and not by science, and in addition this is not a matter for regulation by the ATS but rather by the separate international regime regulating whaling (Watts 1992, 210–11). Likewise, the ever increasing expansion of some scientific research programs in Antarctica have not only raised issues with respect to environmental impact of those operations, but also the possible consequences for territorial claims being ultimately made by the state sponsors of these research activities (De Cesari 1996, 416–18).

Can then Antarctica be used and exploited in the name of science just as other parts of the world have been, or do different principles apply? There has been much debate over the past 50 years over the success of the ATS, however, do these developments create new challenges for the ATS which will increasingly see science pitted against the environment? Indeed, what will be the impact of the current IPY for the next generation of Antarctic science and the next half-century of the ATS? Whilst science is often seen as a form of Antarctic 'currency' (Herr and Hall 1989), it does have the potential to create its own form of conflict. At a time when there is ever increasing attention being given to the polar regions not only for their intrinsic scientific value, but also as their role as sentinels to climate change, there is likewise increased attention being given to the management regimes of the polar regions (Rothwell 1996; Joyner 1998; Tennberg 2000). Whether these regimes are sufficiently robust enough to deal with the challenges ahead is emerging as a key issue (Kriwoken, Jabour and Hemmings 2007; Rayfuse 2007). In Antarctica, where the ATS is so well established and developed, this is an issue of particular significance given the success that has been achieved in setting aside the sovereignty issue for half a century. However, as global oil reserves begin to dwindle and there is increased interest in the resource potential of the polar regions (Christian Science Monitor 2007), the Antarctic regime will inevitably come under pressure. This chapter will address these issues by considering the role of science in the ATS and how science and sovereignty have interacted. The question that will ultimately be posed is whether the ATS is capable of withstanding growing pressures and scientific controversies.

The Antarctic Treaty and the IGY

Whilst the exploration of undiscovered lands and the potential to assert new territorial claims was a significant driver in much of the initial exploration of the Antarctic continent, there was always a strong scientific basis for all Antarctic expeditions in the early part of the twentieth century. Once territorial claims had been asserted to various parts of the continent, science then quickly became the key driver for states to continue to invest in expeditions and bases on the continent and came to represent the only viable 'export' at the time (Laws 1987). As noted elsewhere in this volume (see Jabour and Haward; Woppke), two significant developments occurred in the 1950s which were a direct outgrowth of the ongoing scientific research activity being undertaken on the continent. The first was the establishment by the International Council for Scientific Unions (ICSU) of the 'Special Committee on Antarctic Research' (SCAR), subsequently renamed as the 'Scientific Committee on Antarctic Research', and the second was the decision to hold the IGY in 1957–58. The effect of the IGY was to demonstrate the importance of Antarctic science and the virtue of international cooperation between Antarctic scientists and the twelve countries at that time operating research bases on the continent.[3]

Against the backdrop of a United States proposal in the late 1940s for the internationalization of Antarctica and continuing debate on that topic including within the United Nations (UN) (Hanessian 1960, 435–55; Hawkins 2008), scientists and diplomats saw that an opportunity existed during the IGY process for formal negotiations to commence on the future of the Antarctic continent. From this process, it was hoped that a legal regime could develop which would encapsulate many of the scientific and cooperative goals of the IGY, while also dealing with the underlying political concerns of the seven claimant states[4] and those other states with a strong interest in Antarctic affairs.

This initial promise was confirmed by a series of informal discussions which took place in February and March 1958 during the height of the IGY. The United States at that time put forward a proposal for the development of an Antarctic regime based upon several principles including free access by all nations interested in carrying out scientific research, the growth of scientific cooperation and exchange of information and data amongst participating nations, the use of the continent for peaceful purposes only including non-militarization, guaranteed rights of unilateral access to all parts of Antarctica, and the setting aside of legal status for the duration of the regime so that no one state would need to renounce claims held or in contemplation (Hanessian 1960, 456). Ultimately these principles proved to be pivotal in initial agreement being reached upon the basis for an Antarctic

---

   3   Those countries were Australia, Argentina, Belgium, Chile, France, Japan, New Zealand, Norway, South Africa, United Kingdom, United States and USSR.

   4   Those being Australia, Argentina, Chile, France, New Zealand, Norway, United Kingdom.

regime. In October 1959 the United States hosted a formal diplomatic conference in Washington, and after just six weeks of negotiations the Antarctic Treaty was concluded on 1 December.

The Antarctic Treaty

When compared to some of the very lengthy international instruments which are negotiated today, the Antarctic Treaty seems a very straightforward and even simplistic document. Comprising only fourteen articles, it combines some of the basic measures dealing with demilitarization and conduct of science with some very sophisticated provisions dealing with sovereignty and Treaty review. The Treaty, not surprisingly, focussed on the critical issues of Antarctica's management as identified at the time. Provisions dealing with demilitarization, the importance of science and the resolution of sovereignty claims in particular stand out with all of these matters alluded to in the Treaty's Preamble. Nevertheless, when viewed in a contemporary light, the Treaty has limited provisions dealing with environmental and resource management. This is where the subsequent development of the ATS, and especially the adoption of later instruments dealing with matters such as sealing,[5] marine living resources[6] and comprehensive environmental management[7] has helped to fill in some of these gaps (Vidas 2000).

With the importance of the continuation of scientific research such a motivating factor behind its negotiation, it is unsurprising that such prominence was given to the issue in the Treaty. The Preamble to the Treaty made express reference to the IGY and for the need to continue cooperation towards the 'freedom of scientific investigation' which had developed during that time. The Treaty goes on to provide in Article II:

> Freedom of scientific investigation in Antarctica, and cooperation toward that end, as applied during the International Geophysical Year, shall continue, subject to the provisions of the present Treaty.

From these basic principles, Article III then goes on to provide some further details as to how this objective can be operationalized:

---

5   Convention on the Conservation of Antarctic Seals, done in London, 1 June 1972, in force 11 March 1978 (1972) 11 International Legal Materials 251.

6   Convention for the Conservation of Antarctic Marine Living Resources, done in Canberra, 20 May 1980, in force 7 April 1982 (1980) 19 International Legal Materials 841.

7   Protocol on Environmental Protection to the Antarctic Treaty, done in Madrid, 4 October 1991, in force 14 January 1998 (1991) 30 International Legal Materials 1455.

1. In order to promote international cooperation in scientific investigation in Antarctica, as provided for in Article II of the present Treaty, the Contracting Parties agree that, to the greatest extent feasible and practicable:

(a) information regarding plans for scientific programs in Antarctica shall be exchanged to permit maximum economy and efficiency of operations;

(b) scientific personnel shall be exchanged in Antarctica between expeditions and stations;

(c) scientific observations and results from Antarctica shall be exchanged and made freely available.

Implementation of Article III (1) was further envisaged in Article III (2) through the anticipated establishment of 'cooperative working relations' with relevant UN agencies or other international organizations (Rothwell 2000). These provisions are further supported in Article VIII dealing with the potential inspection of scientific bases and the further exchange of scientific personnel.

There can be no denying that since its adoption, the Treaty has proven invaluable in the continuing promotion of Antarctica as a 'continent of science' in the tradition of the IGY. This has been reflected in various ways such as the US position that the terms of Article II provided for 'free access to all of Antarctica' (United States of America 1960), to the proliferation of scientific bases and stations across the continent such that no part of Antarctica has been sealed off from scientific research. Here it needs to be recalled that the US Amundsen-Scott base at the geographic South Pole straddles the territorial claims made to the continent which converge at that single point. The only real limitation that has been imposed upon the freedom of scientific research throughout the duration of the Treaty has been through additional instruments and mechanisms adopted under the ATS, especially the 1991 Madrid Protocol on Environmental Protection (Bastmeijer 2003, 122–23).

In addition to the important provisions dealing with science, two other aspects of the Treaty need to be mentioned. The first seeks to secure the demilitarization of Antarctica. Article I of the Treaty implements this objective, providing that 'Antarctica shall be used for peaceful purposes only' and goes on to prohibit the conduct of military activities. Consequently, no military bases or fortifications were to be established and military manoeuvres or the testing of weapons were prohibited. An exception was made in the case of military personnel or equipment when used for scientific research or other peaceful purposes. This represented something of an acknowledgement that the scientific activities of a number of Antarctic Treaty parties could not at that time have been conducted without some form of military assistance and indeed there has, in some instances, been a tradition of scientific programs within some countries being directed by the military (Auburn 1982, 95–96). The demilitarization provisions of the Treaty have held fast even when severely tested – as was the case during the 1982 Falklands War when two prominent members of the Treaty, Argentina and the United Kingdom, were engaged in armed conflict. No military operations during the war were ever conducted within the Antarctic Treaty area.

The second aspect of the Antarctic Treaty demanding further attention is Article IV and its provisions dealing with sovereignty. Article IV (1) provides that nothing in the Treaty shall be a basis for an interpretation supporting a renunciation or diminution of previously asserted, existing or even potential claims to Antarctica. This provision thereby sought to deal with the position concerning the existing territorial claims and potential claims that could be made in Antarctica. In doing so, it deals with the interests of a variety of states. These include the seven territorial claimants, those territorial claimants who may be in dispute with other claimants over the validity of their claims, and others such as the United States or Russian Federation (as the successor to the USSR) that may wish to assert a claim in the future. The formula provided is therefore such that all of the principal parties in Antarctic affairs could come together under the control of a single regime without compromising their position on the status of sovereignty claims or potential sovereignty claims (Watts 1992, 127–29).

Article IV (2), on the other hand, deals with other issues arising from sovereignty and which are more directly linked to ongoing scientific research on the continent. To that end, Article IV (2) provides that 'no acts or activities taking place while the present Treaty is in force' shall be a basis for 'asserting, supporting or denying a claim' to sovereignty in Antarctica. At one level, Article IV (2) places limitations on the enhancement of pre-existing territorial claims with the effect that nothing which occurs during the lifetime of the Treaty would further embellish the status of the existing claims. The second aspect creates an outright prohibition on the assertion of new claims or the enlargement of existing claims while the Treaty is in force. The effect of this is therefore that all claims, bases of claims, or potential claims were, in effect, suspended as of the entry into force of the Treaty in 1961 and nothing which occurs while the Treaty is in force will affect the pre-existing position of all of the interested parties – both the claimants and the non-claimants (Rothwell 1996, 75–80; Auburn 1982, 104–10; Triggs 1986, 137–50).

The ATS and science

Once the Antarctic Treaty became operative in 1961 and a new management regime was implemented for Antarctica it was clear that the status quo could not remain in place. Perhaps the most self-evident illustration in the change to how Antarctica was now to be 'governed' was the institution of Antarctic Treaty Consultative Meetings (ATCMs) which were effectively the meeting of states parties to the Treaty. The ATCMs have over the years evolved to increasingly become an essential aspect of the ATS, especially as a result of the additional environmental monitoring provisions created by the 1991 Madrid Protocol, which in turn created the need for additional scrutiny and review of activities having possible environmental impact (Orheim 2000).

However, the Antarctic Treaty created within its provisions a most important criteria for attendance and active participation at the ATCM. Article IX (2) provided that for those states which became a party to the Treaty by way of accession, that to participate at an ATCM a state needed to demonstrate 'its interest in Antarctica by conducting substantial scientific research activity there, such as the establishment of a scientific station or the dispatch of a scientific expedition'. The effect of this provision was that for all those states who became parties to the Treaty after its initial entry into force, which in effect meant any state parties other than the original twelve, and who wished to become active participants at the ATCM, there was a requirement that they demonstrate their bona fides as significant and serious Antarctic research nations.

The interpretation of Article IX (2) became politically contentious when initially there appeared a reluctance on the part of the original Treaty parties to admit any others as consultative parties. As noted by Auburn in 1982:

> Consultative status is the key to the functioning of the Antarctic system. Much of the recent criticism of the Treaty has centred on the heavy burden demanded of new entrants to the decision-making group. Although a number of States have acceded, the granting of Consultative rights has been most restricted lending support to remarks that an exclusive club has been created. It was only after sixteen years that the first new Consultative Party, Poland, was admitted. (Auburn 1982, 147)

During the 1980s, this matter became an issue of significant tension within the wider international community. At that time, the Antarctic Treaty parties were engaged in debates over the merits of developing a minerals regime for Antarctica, which would have facilitated mining on the continent and surrounding Southern Ocean. This opened up the prospect of a possible Antarctic mining boom, and yet there were doubts as to how it would be possible to enjoy the benefits of Antarctic minerals exploitation unless a state was a party to the Treaty and also a consultative party. Whilst becoming a party to the Treaty was simple enough, it was clear from the experience of states like Poland and others that meeting the Article IX (2) criteria for consultative status was very difficult and in many instances beyond the means and capacities of developing states who had little history or experience with polar exploration or science. This argument proved to be particularly forceful in the UN, and the General Assembly began a series of annual debates during the 1980s reassessing the management of Antarctica. Australia, as one of the Antarctic claimants and also a leading Antarctic Treaty proponent, had the role in the UN of leading the response and dealing with the criticism that the Treaty was 'exclusive' and controlled by developed countries (Woolcott 2003, 211).

Eventually there appeared to be a relaxation of the criteria required under Article IX (2) for becoming an Antarctic Treaty consultative party (ATCP), and as a result, there was an increased membership of the Treaty between the mid 1980s and 1990s and accordingly much of the international criticism over the legitimacy of the Treaty evaporated. Nevertheless, central to this debate was the essential criteria that, to be seen as a legitimate member of the ATS, it was necessary for any state to demonstrate its scientific credentials by undertaking significant research activities. While it was arguable that the provisions of Article IX (2) could be subject to legal interpretation (Watts 1992, 15), in reality it was the views of the original Treaty parties – those responsible for negotiating the Treaty in Washington – who counted the most. To that end, it was perhaps inevitable that there would be differences of opinion between those states who had not only been active participants in the IGY but also part of the exploration and scientific discovery of Antarctica since the early twentieth century, and those who had only very recently began to express an interest in polar science.

Another anomaly that arose was that the original twelve members of the Treaty were in effect guaranteed ongoing status as ATCPs effectively as a result of their original interest in Antarctic affairs and science during the IGY and participation at the Washington Conference. Whilst this may have been seen as appropriate given the apparent interests of all the original twelve parties in Antarctic affairs, for two states in particular this was not self-evidently the case. Both Belgium and Japan had some historic interest in Antarctic exploration and science and had been active members of the IGY. Yet, unlike other original parties, they had not asserted territorial claims to the continent, nor did they have strong regional interests (as was the case with South Africa), nor were they scientific or military 'superpowers' (United States and USSR/Russian Federation) who, because of their strategic interests, would have wished to retain a strong interest in Antarctic affairs. Whilst over the years this not proven to be an issue for Japan – which continues to have a strong research presence both on the continent and via its Southern Ocean 'scientific' whaling program – this is not so for Belgium. Over recent years Belgium's interest in Antarctic affairs has waned and its scientific research effort has been considerably reduced, though it seems to have been revived as a result of this fourth IPY (International Council for Science 2007, 69). Nevertheless, it maintains a position of privilege as an ATCP based simply on its being an original party to the Treaty.

A final issue needs to be noted with respect to the significance associated with the provisions of Article IX (2). One of the ways of meeting the criteria associated with demonstrating 'substantial scientific research activity' has been through the establishment of a scientific station. The role of scientific stations in Antarctica is of immense importance to the history of human activities in Antarctica, including long abandoned stations, which in some instances have been preserved as historic sites. One example is the case of Mawson's Hut in the Australian Antarctic Territory (Australian Antarctic Division 2008; Garrett 2008). In other instances where stations have been abandoned, as was the case with a number of bases which were deemed unsustainable following the conclusion of the IGY or as a result of the rationalization of Antarctic scientific programs, issues have arisen with respect to environmental impact and how those bases should be subject to clean-up and removal of hazardous wastes. This has become an even more pressing issue since the introduction of the 1991 Madrid Protocol (Bastmeijer 2003, 274–75). However, in addition to the issue of the pre-1959 scientific bases there is a question as to how many scientific bases are sustainable in Antarctica and how states wishing to meet the criteria for ATCP status under Article IX (2) are to go about the business of establishing a scientific research station. This is especially prominent in the contemporary era when all such activity would be subject to very rigorous environmental impact assessment under the Madrid Protocol. One impact that arose from the scramble of some states to assert their scientific credentials was the proliferation of research bases over parts of the Antarctic Peninsula and its adjacent islands. This region of the continent offered the advantage of being relatively ice free, enjoyed more temperate climates, and because of its proximity to South America was more accessible than many other parts of the continent. However, this resulted in serious concerns being raised as to environmental impact. Many of these matters have now been addressed as a result of the implementation of the Madrid Protocol and a common sense approach is being taken in the sharing of scientific bases which in turn has seen the ATCM adopt more streamlined procedures for acceptance of new ATCPs based on the current realities of undertaking 'science' in Antarctica. These issues highlight not only the political sensitivity associated with the conduct of science in Antarctica, which have arisen even under the terms of the Treaty, but also the environmental dimensions of the activity. Currently there are 53 scientific research facilities in Antarctica operated by 27 national Antarctic programs[8] (Council of Managers of National Antarctic Programs 2008), of which two facilities are shared between two national programs.

The 2007–08 IPY

The 2007–08 IPY has clear connections with the IGY which, as has been noted, provided the foundation upon which the Antarctic Treaty and the subsequent ATS were ultimately based. To that end, the IPY not only seeks to advance polar science, but it also ultimately reinforces and even celebrates the importance of science in Antarctica. Scientists from 63 nations were part of the IPY, which included a total of 44 Antarctic and 65 Bipolar research projects. This offers a sense of the scope of the IPY and how it is very much continuing in the historical traditions of the IGY (International Council for Science 2007). The emphasis of an IPY legacy, seeking to ensure that infrastructure and polar observing systems are developed with a particular focus upon environmental baseline studies (International Council for Science 2007, 15), also reinforces some of the roots of the original IGY which was

---

8   The national programs being Argentina, Australia, Brazil, Bulgaria, Chile, China, Ecuador, Finland, France, Germany, India, Italy, Japan, New Zealand, Norway, Peru, Poland, Romania, Russia, South Africa, South Korea, Spain, Sweden, United States, United Kingdom, Ukraine, Uruguay.

then reinforced by Article III of the Antarctic Treaty. Yet, the IPY has bowed to the impact of the ATS, and unlike its earlier version, there has been no expansion of existing Antarctic scientific research bases and stations, inevitably due to the very significant limitations now created by the Madrid Protocol on the building of such facilities in Antarctica and the growing awareness and concerns over environmental impact.

Southern Ocean bioprospecting

Since the early part of this century there has been increasing interest in bioprospecting in Antarctica and most particularly the Southern Ocean. More so than the continent, the Southern Ocean presents greater prospects for new discoveries because of the abundance of marine life and still relatively unexplored biological domain (Jabour-Green and Nicol 2003). In this respect, the resource component of bioprospecting reaffirms why as Joyner has argued science has become so central to human activity and governance of Antarctica (Joyner 1998, 186).

Bioprospecting can be defined as:

> the search for and sourcing of organisms from the natural environment with the purpose of extraction of compounds for further investigation of their potential for development in therapeutic or industrial applications. (Jones 1998, 89)

As noted by Rogan-Finnemore, bioprospecting is a relatively recent activity with the primary focus being the search for 'novel biodiversity, whose component parts may then be utilized in a product or process and developed for commercialization' (Rogan-Finnemore 2004, 3). At one level, it could be argued that this new phenomena is yet another challenge to the Antarctic legal regime requiring a swift response. In the alternate, it can be argued that there already exists a sufficiently comprehensive legal regime in place in the Southern Ocean to address this new challenge. This view can be supported not only on the basis of the UN Law of the Sea Convention (UNCLOS) and the specific framework of the ATS, but also through recent developments in international environmental law, in particular the 1992 Convention on Biological Diversity (CBD) (Jabour-Green and Nicol 2003, 94–97).[9]

The relatively recent development of bioprospecting has created challenges for its legal regulation (Jones 1998). The question remains as to whether or not this is an activity which is best regulated under national laws, or is appropriate for international legal regulation. If international responses are adopted, are general framework provisions in order or is there a need for detailed laws? If so, over which areas should these laws apply – the high seas only, or would it extend to

---

9   (1992) 31 International Legal Materials 818, in force 29 December 1993.

both the oceans and terrestrial areas? To understand how bioprospecting can take place under the legal regime of the Southern Ocean it is therefore necessary to refer to the law of the sea. Much of the law of the sea is directed towards confirming and clarifying the extent of coastal state sovereign rights and jurisdiction over adjacent maritime areas. Starting with the territorial sea, over which coastal states have almost exclusive sovereign rights and jurisdiction, the law of the sea recognizes a mix of sovereign rights and jurisdiction over adjacent maritime areas up to the outer edge of the continental shelf. Particularly important in this respect is the recognition of resource sovereignty over the water column, the sea-bed and subsoil up to the limit of the 200 nautical mile exclusive economic zone (EEZ).[10] Continental shelf sovereign rights over the sea-bed and subsoil which extend to the exploration and exploitation of natural resources can be projected in the instance of extended continental shelf claims up to 350 nautical miles from the coast.[11]

Beyond the limits of the continental shelf/EEZ are two remaining maritime zones. The first is the high seas, which is the water column including the surface beyond the limit of the 200 nautical mile EEZ. High seas freedoms are some of the most historic rights recognized under international law and traditionally extend to freedom of fishing and freedom of navigation. Under the modern law of the sea, it also extends to the freedom to conduct marine scientific research.[12] The other regime is that dealing with the deep sea-bed, which is the 'Area' beyond the limits of national continental shelf claims.[13] Regulated under Part XI of the UNCLOS by a regime designed to recognize the so-called 'common heritage of mankind' principle, the resources of the Area are beyond the sovereign control of any one state and are regulated in conformity with the Convention's provisions by the International Sea-Bed Authority.

Pivotal to the law of the sea regime dealing with bioprospecting is Part XIII of the UNCLOS Convention dealing with marine scientific research. Part XIII makes clear that all states have a right to conduct marine scientific research subject to the rights and duties of other states, including coastal states.[14] The framework which is then constructed, is as follows:

- within the Territorial Sea the coastal state has the exclusive right to regulate marine scientific research meaning that the activity can be either prohibited, or permitted subject to regulation;[15]

---

10   UNCLOS, art. 56.

11   UNCLOS, arts. 76 and 77.

12   UNCLOS, art. 87.

13   UNCLOS, art. 1 (1) defines the 'Area' as the sea-bed and ocean floor and subsoil thereof, beyond the limits of national jurisdiction.

14   UNCLOS, art. 238.

15   UNCLOS, art. 245.

- within the EEZ and continental shelf the coastal state can regulate marine scientific research on the basis that in normal circumstances consent will be granted for such research to be undertaken by other states;[16]
- in the Area and within the water column beyond the 200 nautical mile limit which would in most cases be high seas, all states have the right to conduct marine scientific research.[17]

Taking into account this framework and the provisions of the law of the sea and ATS as they apply to the Southern Ocean, the following can be concluded:

1. That within the Southern Ocean, Antarctic claimant states have a capacity to proclaim maritime zones from the continent and adjacent islands;
2. That Antarctic claimant states also have the capacity to explore and exploit the water column and sea-bed;
3. That Antarctic claimant states can legitimately seek to determine the outer limits of their continental shelf claims;
4. That Antarctic claimant states can likewise seek to regulate marine scientific research within their adjacent territorial sea and EEZ/continental shelf areas;
5. That parts of the Southern Ocean may be considered to be part of the 'Area' and fall under the Part XI deep sea-bed regime;
6. That the extent of maritime claims off the Antarctic continent need to be reconciled with:
   a. The provisions of the Antarctic Treaty, article IV, which place limitations upon the assertion of new claims
   b. That a sector of the Antarctic continent has not been subject to recognized claim and as a result there is no current potential for maritime zones to be generated offshore that area.

Given the unique legal regime that exists in the Southern Ocean, combined with sovereignty claims which are legally and politically constrained by the ATS, a legal regime for bioprospecting in the Southern Ocean could revolve around the following key actors. Firstly, coastal states will play a key role as recognized by the marine scientific research provisions of Part XIII of the UNCLOS. They have a clear-cut capacity to regulate bioprospecting within the narrow twelve nautical mile territorial sea, and further beyond to the limits of the EEZ and continental shelf. Secondly, flag states play an important role in regulating the activities of their ships engaging in marine scientific research to ensure compliance not only with the general provisions of the UNCLOS but also with the CBD. In addition, flag states also have an important role in ensuring their ships meet the stringent marine environmental provisions which apply within the Southern Ocean. Thirdly,

---

16   UNCLOS, art. 246.
17   UNCLOS, arts. 256–57.

there is also a role for the international community more generally, especially as envisaged under the CBD but also under the provisions of the UNCLOS dealing with the deep sea-bed and the regulation of activities in the 'Area' by the International Sea-Bed Authority.

However, when the legal regime of the ATS is combined with the political history of the region and mixed with the 'culture' of promoting the freedom of scientific research, there remains legal uncertainty as to how bioprospecting will be adequately regulated in the Southern Ocean. To begin, the status of the seven territorial claims to the Antarctic continent remains uncertain. Few states outside of the original members of the Antarctic Treaty have actively recognized the seven claims to the continent. In addition, the capacity of the claimant states to assert maritime claims has also been questioned because of the Treaty's prohibition in Article IV on the assertion of a 'new claim, or the enlargement of an existing claim'. This reflects the reality of the situation with nearly all of the claimant states retaining an ambiguous position on their various claims to a territorial sea, EEZ, or a continental shelf (Kaye and Rothwell 1995). In addition, Article VIII of the Antarctic Treaty also places limitations on the active assertion of national jurisdiction within the Antarctic Treaty area, which further compromises the capacity to regulate bioprospecting conducted by scientific expeditions from other Antarctic Treaty parties.

It could be anticipated that flag state enforcement of bioprospecting laws and regulations would be an adequate alternative, however this also does not present any guarantee for regulation or control of bioprospecting. Given that marine bioprospecting substantially falls under the regulation and control of coastal states within their maritime zones, if flag states do not recognize coastal state sovereignty and jurisdiction but rather consider the Southern Ocean to substantially be high seas and therefore open to all forms of marine scientific research, then the capacity to regulate ships engaged in bioprospecting is further compromised. Indeed, the view could be taken that through a combination of the UNCLOS and the CBD, much of the Southern Ocean could be considered open to bioprospecting and accordingly only subject to the most basic of controls. On this scenario it could be anticipated that only ATS member states would respect a distinctive Southern Ocean regime regulating bioprospecting.

Further difficulties arise because of the global legal regime created under the CBD and the UNCLOS which promote both marine scientific research and bioprospecting. It could even be argued that given the potential global benefits arising from bioprospecting, such an activity is also generally consistent with the 'common heritage' concept as it applies not only to the deep sea-bed but also to other 'commons' such as Outer Space and the Moon (Joyner 1998, 222–35). When viewed against the ATS, there is a clear tension between the global regime and the regional as there is currently no framework within the Antarctic regime to adequately regulate bioprospecting. Any attempt by the claimant coastal states to regulate the activity could be rejected outright by a large number of other states who reject the legitimacy of the ATS and territorial claims which have been made.

Scientific research and Japanese whaling in the Southern Ocean

One issue concerning the conduct of scientific research in Antarctica that has generated considerable debate in recent years has been Japan's conduct of scientific research into whales in the Southern Ocean (Ackerman 2002). This research program, which now has a twenty-year history stretching back to the late 1980s (Darby 2007), at one level falls within the ambit of the 1946 International Convention for the Regulation of Whaling (ICRW),[18] but may also be seen as falling within the scope of the ATS. The ICRW was originally an instrument that sought to promote commercial whaling. However, over time it has evolved into a regime with a strong focus on the conservation of whales. This is currently reflected in the ICRW Schedule, para. 6, which introduced a moratorium on commercial whaling from 1986. The ICRW, however, does not prohibit the killing or taking of all whales. Article VIII of the Convention permits state parties to issue special permits authorizing the taking and killing of whales for scientific purposes. State parties issuing permits under Article VIII are required to report to the IWC – the Commission established as the Secretariat for the ICRW – on all authorizations granted for so-called 'scientific whaling'.

Japan introduced the Japanese Whale Research Program under Special Permit in the Antarctic (JARPA) in the 1987–88 season and continued that program until the 2004–05 season. JARPA had a principal focus upon research into minke whales in the Southern Ocean with initially a sample size of 300 (plus or minus ten per cent) being taken each season for research purposes. Beginning with the 1995–96 season the sample size was increased to 400 (plus or minus ten per cent). In 2005, Japan announced its intention to conduct the second phase of the Japanese Whale Research Program under Special Permit in the Antarctic (JARPA II) with projected annual sample sizes being 850 minke whales, 50 humpback whales, and 50 fin whales (plus or minus ten per cent allowance) (Institute of Cetacean Research 2008a). The JARPA II research program by Japan in the Southern Ocean has generated considerable controversy, especially with respect to the legitimacy of the research program under the ICRW. At the 2005 IWC Annual Meeting Resolution 2005–1 (30 votes for, 27 against, 1 abstention) was adopted which urged 'the Government of Japan to withdraw its JARPA II proposal or to revise it so that any information needed to meet the stated objectives of the proposal is obtained using non-lethal methods'. Japan ignored this request. Likewise, at the 59th meeting of the IWC in Anchorage in 2007, a further resolution was adopted which called upon Japan to address recommendations by the IWC Scientific Committee relating to the JARPA I program and 'to suspend indefinitely the lethal aspects of JARPA II conducted within the Southern Ocean Whale Sanctuary'.[19] Japan has ignored these calls to halt its scientific research program. As a result, Australia, a party

---

18   [1948] Australian Treaty Series No. 18.

19   Resolution 2007–1, 59th Annual Meeting of the International Whaling Commission, Anchorage, Alaska, USA, 2007.

to the ICRW, has suggested that it may seek to challenge the legitimacy of the Japanese whaling program before international courts (Smith and Garrett 2007). A significant legal issue that arises here is whether Japan's conduct of JARPA II is consistent with not only the ICRW but also the provisions of the ATS. Can Japan, for example, point to the Antarctic Treaty's promotion of the freedom of scientific research to argue that its activities cannot be subject to regulation and that no other country has the capacity to contest the legitimacy of its activities?

First, with respect to the ICRW, Japan contends that the conduct of 'scientific whaling' is consistent with the provisions of Article VIII of the Convention (Institute of Cetacean Research 2008b). Article VIII gives to any ICRW party an extensive capacity to issue special permits for the taking or treating of whales for the purposes of scientific research. It is made clear that this provision represents an exemption from the other provisions of the ICRW. This is also reflected in the minimal obligations imposed upon contracting parties which grant such permits. With the exception of the annual transmittal of scientific data originating from a scientific research program, there are few constraints imposed upon a contracting government issuing permits to its nationals to conduct a scientific research program. However, there is considerable scope for varying interpretations as to what constitutes 'science' for these purposes (Harris 2005), especially when there is evidence that Japan's research program has a commercial element to it (Palmer 2007), not to mention emerging evidence as to the capacity to conduct non-lethal scientific research into whale stocks (Australian Antarctic Division 2005).

While the conduct of scientific research into whales under the Convention is accorded special status it does not follow that there are absolutely no constraints upon the exercise of a right of scientific research. In general international law, the notion of abuse of right recognizes that a state may not exercise a right in a way which impedes the enjoyment by other states of their own rights, or exercise that right for an end different from that for which the right was created to the injury of another state (Triggs 2000, 42). Setting aside the provisions of the ICRW, what does the ATS have to say about Japan's scientific research activities in the Southern Ocean? To begin with it needs to be noted that the Antarctic Treaty extends well beyond the continent up to the limits of the surrounding oceans to 60° S and accordingly there is no doubt as to the capacity of the Treaty and the ATS more generally to operate within the waters of the Southern Ocean. However, Article VI of the Treaty provides for the retention of rights under international law as they apply to areas of high seas within this area. At the time of the conclusion of the Treaty and its eventual entry into force in 1961, other than for very narrow areas of territorial sea claims, overwhelmingly the vast expanses of the Southern Ocean would have been considered high seas (Auburn 1982, 129–38; Watts 1992, 149). Whilst the rights associated with marine scientific research were relatively undefined at that time, the modern law of the sea makes it very clear that one of the freedoms of the high seas is the capacity to undertake marine scientific research.[20]

---

20   UNCLOS, Article 87.

At one level therefore, it would seem there is a correlation between not only the law of the sea but also the Antarctic Treaty with respect to the ability to conduct scientific research into whales in the Southern Ocean.

However, this is an area where the provisions of the 1991 Madrid Protocol may have application. The Protocol sought to extend the provisions of the Antarctic Treaty with respect to environmental protection in the Antarctic Treaty area. To that end, Article 3 of the Protocol lists a number of environmental principles for the protection of the Antarctic environment. These principles extend to a range of activities, including scientific research programs and tourism which are to take place in a manner consistent with the Protocol and be modified or suspended if they threaten to result in impacts upon the Antarctic environment (Bastmeijer 2003, 123). This suggests a possible conflict between the Protocol and the ICRW with respect to scientific research, however the Protocol seeks to address issues of consistency with other international instruments operative within the ATS area and Article 5 provides that the state parties are to 'consult and co-operate' when there are multiple international instruments operating within the same area so as to try and avoid 'inconsistency' between the implementation of those instruments and the Protocol. Consultation should therefore be taking place between state parties to the Protocol and parties to other international instruments operative within the Southern Ocean in order to achieve as much as possible a harmonization of objectives. On this interpretation, it would be expected that there would be consultation between the state parties to the two regimes – the ATS and the ICRW – to discuss how two apparently conflicting provisions – one permitting unilateral scientific whaling and the other placing constraints upon activities that may have significant environment impact – can possibly be harmonized. However, to date, no such consultations have occurred.

Ultimately, the manner in which Japan's conduct of JARPA II has been undertaken exclusive of the ATS regime seems difficult to reconcile with the legal regime contemplated under the Protocol. The stated intent of JARPA II relates to matters which clearly fall under the mandate of the Protocol because of a geographical connection with the Southern Ocean and the fact that the research program is to be undertaken by Japan, who is also a party to the Protocol. JARPA II is also a matter which clearly falls under the mandate of the ICRW. Accordingly, this mutual interest in JARPA II should be sufficient to activate the provisions of Article 5 of the Protocol so as to at a minimum facilitate communications and exchanges of views between state parties to both instruments but also the respective international institutions. The undertaking of JARPA II without any form of environmental impact assessment (EIA), notwithstanding the scientific basis for the program, is contrary to the environmental principles of the Protocol. In order to avoid an inconsistency between any right Japan may claim under the ICRW and the Madrid Protocol it is therefore necessary for Japan to submit JARPA II to an environmental impact assessment under the provisions of the Madrid Protocol.

Concluding remarks

Antarctica truly remains a unique continent because of the enduring significance that is associated with scientific research, historically embedded in the conduct of activities on the continent and surrounding ocean by the various polar years, highlighted by the IGY which in turn provided a catalyst for the negotiation of the Antarctic Treaty which sought to perpetuate the spirit of the IGY by seeking to ensure the ongoing freedom of scientific research, and which in the twenty-first century has again been reinforced by the IPY. This blending of science, law, politics and community which is found nowhere else on the planet has in recent years been reinvigorated by the global focus on climate change and an awareness of the importance of the polar regions as sentinels in climate research. Hence, science and Antarctica have to an extent turned the full circle and the science of the new millennium is proving to be just as important as that conducted in the last century when polar research was a relatively new discipline and Antarctica in particular was just in the process of being opened to wide ranging scientific endeavour. Yet while the conduct of Antarctic science within an enduring framework which has permitted a great cooperative spirit by and large free of politics is to be rightly celebrated, this should not suggest that Antarctic science remains free of controversy. The conduct of bioprospecting holds the seeds of possible dispute, especially if distinctions are sought to be drawn between the freedoms associated with the exploitation of genetic resources and the limitations on natural resource exploitation more generally, especially Antarctic minerals. This is highlighted, though in a very different sense, by the conduct of 'scientific' whaling in the Southern Ocean. Japan's ongoing whaling program under the ICRW attracts considerable international controversy outside of the Antarctic regime and yet, is apparently ignored within the ATS including from some of the Japan's fiercest anti-whaling critics such as Australia, New Zealand and the United Kingdom. Whilst this contradiction is difficult to reconcile, it does highlight the great lengths to which members of the ATS will go to in an effort to effectively 'quarantine' certain issues so that they do not impact upon the harmony of the regime. This has facilitated the conduct of much Antarctic science during the past fifty years and on the evidence to date may well be capable of standing the test of time long into the future.

References

Ackerman, R.B. (2002), 'Japanese Whaling in the Pacific Ocean: Defiance of International Whaling Norms in the Name of "Scientific Research", Culture and Tradition', *Boston College International and Comparative Law Review* 25, 323–41.

Auburn, F. (1982), *Antarctic Law and Politics* (London: Croom-Helm).

Australian Antarctic Division (2005), 'Australian Announces Whale Conservation Initiatives' (28 October 2005). Hobart: Australian Antarctic Division at <www.aad.gov.au/default.asp?casid=20971>, accessed 10 June 2008.

—— (2008), 'Conservation of Mawson's Huts'. Hobart: Australian Antarctic Division at <www.aad.gov.au/default.asp?casid=12153>, accessed 10 June 2008.

Bastmeijer, K. (2003), *The Antarctic Environmental Protocol and its Domestic Legal Implementation* (The Hague: Kluwer Law International).

Blay, S. and Bubnna-Litic, K. (2006), 'The Interplay of International Law and Domestic Law: The Case of Australia's Efforts to Protect Whales', *Environmental and Planning Law Journal* 23, 465.

Bush, W.M. (ed.) (1988), *Antarctica and International Law: A Collection of Inter-State and National Documents* (London: Oceana).

Christian Science Monitor (2007), 'Editorial: Scramble for the Arctic', *Christian Science Monitor* 21 August 2007 <www.csmonitor.com/2007/0821/p08s01-comv.html>, accessed 10 June 2008.

Council of Managers of National Antarctic Programs (2008), 'Antarctic Facilities in Operation' at <www.conmap.aq/operations/facilities> , accessed 7 July 2008.

Darby, A. (2007), *Harpoon: Into the Heart of Whaling* (Crows Nest, NSW: Allen & Unwin).

De Cesari, P. (1996), 'Scientific Research in Antarctica: New Developments', in Francioni and Scovazzi (eds).

Francioni, F. and Scovazzi, T. (eds) (1996), *International Law for Antarctica,* 2nd edition (The Hague: Kluwer Law International).

Garrett, P. (2008), 'Plan to Protect Antarctic Home of Australian Pioneer' Australian Government – Media Release (18 June 2008) at <www.environment.gov.au/minister/garrett/2008/pubs/mr20080618.pdf>, accessed 10 July 2008.

Handmer, J.W. (ed.) (1989), *Antarctica: Policies and Policy Development* (Canberra: Centre for Resource and Environmental Studies, Australian National University).

Hanessian, J. (1960), 'The Antarctic Treaty 1959', *International and Comparative Law Quarterly* 9, 436–80.

Harris, A.W. (2005), 'The Best Scientific Evidence Available: The Whaling Moratorium and Divergent Interpretations of Science', *William and Mary Environmental Law and Policy Review* 29, 375–50.

Hawkins, A. (2008), 'Defending Polar Empire: Opposition to India's Proposal to Raise the "Antarctic Question" at the United Nations in 1956', *Polar Record* 44:228, 35–44.

Hayton, R.D. (1960), 'The Antarctic Settlement of 1959', *American Journal of International Law* 54, 349–71.

Hemmings, A. and Rogan-Finnemore, M. (2004), *Antarctic Bioprospecting.* Gateway Antarctica Special Publication Series #0501 (Christchurch: University of Canterbury).

Herr, R. and Hall, R. (1989), 'Science as Currency and the Currency of Science', in Handmer (ed.).

Herr, R.A., Hall, H.R. and Haward, M.G. (eds) (1990), *Antarctica's Future: Continuity or Change?* (Hobart: Australian Institute of International Affairs).

Institute of Cetacean Research (2008a), 'JARPA II: The Second Phase of Japan's Whale Research Program under Special Permit in the Antarctic', Tokyo: Institute of Cetacean Research at <www.icrwhale.org/FAQ.htm>, accessed 10 July 2008.

—— (2008b), 'Japan's Whale Research in the Antarctic – Backgrounder', Tokyo: Institute for Cetacean Research at <www.icrwhale.org/eng/background.pdf>, accessed 10 July 2008.

International Council for Science (2007), *The Scope of Science for the International Polar Year 2007–2008* (Geneva: World Meteorological Organization).

International Polar Year (2008), 'Ideas for IPY: National' IPY 2007–2008 at <classic.ipy.org/development/ideas/national.index.htm>, accessed 10 July 2008.

Jabour-Green, J. and Haward, M. (eds) (2001), *The Antarctic: Past, Present and Future*. Hobart: Antarctic CRC Research Report # 28. (Hobart: University of Tasmania).

—— and Nicol, D. (2003), 'Bioprospecting in Areas Outside National Jurisdiction: Antarctica and the Southern Ocean', *Melbourne Journal of International Law* 4, 76.

Jones, J.S. (1998), 'Regulating Access to Biological and Genetic Resources in Australia: A Case Study of Bioprospecting in Queensland', *Australasian Journal of Natural Resources Law and Policy* 5, 89.

Joyner, C.C. (1998), *Governing the Frozen Commons: The Antarctic Regime and Environmental Protection* (Columbia, South Carolina: University of South Carolina Press).

Kaye, S. and Rothwell, D.R. (1995), 'Australia's Antarctic Maritime Claims and Boundaries', *Ocean Development and International Law* 26, 195.

Kriwoken, L.K., Jabour, J. and Hemmings, A.D. (eds) (2007), *Looking South: Australia's Antarctic Agenda* (Annandale: The Federation Press).

Laws, R.M. (1987), 'Scientific Opportunities in Antarctica', in Triggs (ed.).

Orheim, O. (2000), 'The Committee on Environmental Protection: Its Establishment, Operation and Role within the Antarctic Treaty System', in Vidas, D. (ed.)

Palmer, G. (2007), 'Is This Just Commercial Whaling in Drag?', *The Environmental Forum* 24:3, 45.

Rayfuse, R. (2007), 'Melting Moments: The Future of Polar Oceans Governance in a Warming World', *Review of European Community and International Environmental Law* 16, 196–216.

Rogan-Finnemore, M. (2004), 'Setting the Scene', in Hemmings and Rogan-Finnemore (eds).

Rothwell, D.R. (1996), *The Polar Regions and the Development of International Law* (Cambridge: Cambridge University Press).

—— (2000), 'Relationship between the Environmental Protocol and UNEP Instruments', in Vidas (ed.).

—— and Davis, R. (1997), *Antarctic Environmental Protection: A Collection of Australian and International Instruments* (Annandale: The Federation Press).

Smith, S. and Garrett, P. (2007), 'Australia Acts to Stop Whaling', Australian Government – Joint Media Release (19 December 2007) at <www.foreignminister.gov.au/releases/2007/fa-s002_07.html>, accessed 10 July 2008.

Tennberg, M. (2000), *Arctic Environmental Cooperation: A Study in Governmentality* (Aldershot: Ashgate).

Triggs, G. (1986), *International Law and Australian Sovereignty in Antarctica* (Sydney: Legal Books).

—— (ed.) (1987), *The Antarctic Treaty Regime* (Cambridge: Cambridge University Press).

—— (2000), 'Japanese Scientific Whaling: An Abuse of Right or Optimum Utilization?', *Asia Pacific Journal of Envi onmental Law* 5, 33–59.

United States of America (1960), 'Senate Report on Grounds for Opposing Ratification of the Antarctic Treaty', reprinted in Bush (ed.).

Vidas, D. (ed.) (2000), *Implementing the Environmental Protection Regime for the Antarctic* (Cambridge: Cambridge University Press).

Wansborough, T. (2004) 'On the Issue of Scientific Whaling: Does the Majority Rule?', *Review of European Community and International Environmental Law* 13, 333.

Watts, A. (1992), *International Law and the Antarctic Treaty System* (Cambridge: Grotius Publications).

Woolcott, R. (2003), *The Hot Seat: Reflections on Diplomacy from Stalin's Death to the Bali Bombings* (Sydney: Harper Collins).

Chapter 7

The Formation and Context of the Chilean Antarctic Mentality from the Colonial Era through the IGY

Consuelo León Woppke

Introduction

Scholars have analyzed the formation of the Antarctic Treaty of 1959 from a variety of perspectives, often emphasizing how its theme of peaceful scientific cooperation was overshadowed by Cold War maneuvering. It has recently been demonstrated that, at the time, some US officials viewed the treaty as potentially detrimental to their nation's strategic interests, for which reason the government made no concerted effort to encourage its ratification by the senate (Moore 2008). This geopolitically oriented perspective has also suffused recent literature presenting the Antarctic as a component of US–Chilean relations. The smaller American republic has been credited with exerting a much greater influence on the formation of the Treaty than seemed likely given its size and relative insignificance on the global stage (for example, Moore 2001, 2003).

This chapter highlights the influence of domestic political and scientific factors on Chile's contribution to the formation of the Antarctic Treaty. It provides a synthesis of diverse materials, many of which either have not been previously cited or have not been presented within the context of English-language scholarship. Though the emphasis lies primarily on the 1940s and 1950s, the period in which two US-sponsored international proposals were debated, the origin of Chile's Antarctic 'mentality' is traced back to the colonial era. The traditional historical methodology of correlating primary and secondary sources is followed with great attention to detail and willingness to 'let the facts speak for themselves' rather than to portray them as evidence of any preconceived theory or judgment.

Chilean interest in Antarctica in the nineteenth century

Europeans first sighted what is today known as Chile in November 1520 while Hernando de Magallanes was leading a voyage around the world, passing through the straits later named in his honor. The explorers correctly speculated that the straits would provide an ideal point of departure for reaching the continent believed to lie at the bottom of the world, which at the time was referred to as *Terra Australis Incognita*. Yet the frozen continent would not be sighted for approximately three more centuries, and Chile would not mount an Antarctic expedition for approximately another century thereafter. In the interim Chile's leaders consistently sought to protect its Antarctic rights, based on geographical contiguity, a royal treaty, a papal decree (Moore 2003, 69), and the Latin American principle of mutual territorial recognition (*uti possidetis*) from the tip of South America to the bottom of the world. Otto Nordenskjöld was the first to speculate about the link between the Andes and the Antarctic peninsula (Pacheco 1930; Martinic 1999).

In the nineteenth century few Chileans took interest in the Antarctic and, unlike the Spanish, they rarely pondered the value of sending expeditions there ('La Exploración' 1998). In part this was because no other nation contested Chileans' Antarctic rights, which reinforced their belief that, by default, those rights extended to as-yet-undiscovered territory (Vicuña 1915). Another factor was the nation's skeptical attitude toward science, the pursuit of which might have convinced officials to mount an expedition, even if they had questioned the others benefits of doing so (Drake 1978). In the realm of imperial politics, many Europeans regarded scientific knowledge as an effective means of consolidating or expanding state power.[1] While some Chileans must have shared this attitude, their nation was unable to project its power on a global level, and its citizens had been conditioned to view science as inconsistent with their faith in Christian dogma, especially as promulgated by the Catholic Church. It is perhaps unsurprising that, as a result, most scientists were either newly arrived foreigners hired by the government or freethinkers, which never fully accepted the prevailing social beliefs (Gutierrez and Gutierrez 2006). This fact, coupled with the nation's geographical isolation and the non-disclosure of many European scientific advances, explained why Chileans remained largely ignorant of the voyages and discoveries of Protestant navigators (Kirwan 2001, 88). Nonetheless, during the earliest decades of the Chilean republic, many citizens grew to appreciate the success of British and US sealers and whalers who were exploiting the sub-Antarctic waters (Evans 1957, 25; Vercel 1942, 23). The first US explorer believed to have reached the Antarctic Peninsula was Nathaniel Palmer in 1821 (Joyner and Theis 1997, 21).

There is no evidence that Chileans were aware that in 1835 the British Association of Dublin implored London to send any expeditions to the southern polar region. At the time many Latin American nations were distracted by the need to stabilize their political systems, and Chile was no exception. Not until the following decade were its institutions sufficiently viable to support an expedition

---

1   During the colonial era, most Spanish and Chilean scientists were at least partially affiliated with the Catholic Church. For example, Spanish scientists Manuel de Ulloa and Jorge Juan, who made valuable contributions in relation to data collected in sub-Antarctic waters, were affiliated with the armed services, the aristocracy, and the church (Soler 2002).

to assess the feasibility of colonizing the Straits of Magellan and the Antarctic territory which lay below. During the same decade, the government established the nation's first secular institution of higher education, the University of Chile. Its academics closely followed or hoped to follow the ongoing, scientific developments in Europe and the United States. However, these academics represented a small minority of the population whose viewpoints often were at odds with those of the Catholic Church, which enjoyed much greater political and educational influence (Domeyko 1847). As such, Chilean science largely stagnated. In 1882, Miguel Luis Amunátegui, secretary general of the University of Chile, lamented the inability of scientists to attract the degree of public funding which their research required in order to be truly first rate (Zegers 1876).

The so-called Movement of 1848 revived Europeans' devotion to free thought and empirical research; two of the ideals most closely associated with the French Enlightenment. This trend failed to generate widespread enthusiasm in Chile, though it did inspire the anti-clerical elite that formed the nucleus of the University of Chile and the Liberal Party. Meanwhile, European expeditions to the Pacific and Southern Oceans continued; some of them reaching the Antarctic peninsula and greatly advancing global knowledge of the coastal cartography. Again, only a limited amount of this knowledge reached the Chilean scientific community, failing to alter the mindset of those individuals responsible for funding the community's development.

Global interest in the Antarctic rose in the 1870s, a decade during which several scientific congresses were held. Institutions around the world devoted funds to explore issues such as magnetism, meteorology, and auroras. Chile failed to join the global efforts primarily due to its involvement in the War of the Pacific (1879–83) against its Peruvian and Bolivian neighbors. Nevertheless, Chile did not remain completely oblivious of developments.

The Chilean Astronomic Service, based in Santiago, and the local government of Punta Arenas provided assistance to the expeditions of Adrien de Gerlache, Roald Amundsen, Jean Batista Charcot, Robert F. Scott, Erich von Drygalsi and Ernest Shackleton (Cook 1899; Berguño and Canales 2005; Martinic 1982). Chilean assistance also helped to sustain the expeditions of Otto Nordenskjöld. In 1875, the newly created Hydrographical Office of the Chilean Navy also began to assist other nations' quest to decipher the untold mysteries of the Antarctic though refraining from leading any expeditions of its own (Martinic 1982). The Chilean Naval Attaché in London Luís A. Goñi attended European geographical conferences (Escudero 1953).

During the First Polar Year (1882–83), international cooperation substantiated hope that data obtained from various stations could be shared and effectively correlated to advance knowledge of the continent formerly known as *Terra Incognita Australis* ('Temporary Commission' 1950). During this event, Chile's scientific activities were directed by institutions affiliated with the government or the armed services, for example, the Hydrographic Office of the Navy, created in 1874, the Seismologic and Meteorologic Services, the Chilean Society of History and Geography, and the Scientific Society (Berguño 1999). Ninety days after x-rays were discovered in Europe, they were reproduced by two professors at the University of Chile, as indicated in an 1896 publication by the Chilean Scientific Society. Carlos Moesta was Chile's representative to the Polar Year. Various historians have maintained that the non-development of Chilean science after the Pacific War (1879–1884) is simply a 'myth' (Gutierrez and Guitierrez 2006). It is true that there was some development in specific areas, such as physics, due to the obvious military and economic value, yet there remains little evidence to suggest that the nation's devotion to science was permanent. Naval personnel attempted to keep abreast of scientific discoveries by attending international congresses. In 1884 Alejandro Bertrand published his influential map which was the first to 'illustrate the geographical basis for an Antarctic sovereignty claim' which in turn initiated the formation of Chile's south polar consciousness (Berguño 1999; López 1975).

In the late nineteenth century, the studies of Rodolfo Philippi, director of the Chilean Natural History Museum, and Federico Alberto of the Ministry of Industry's Office of Zoological and Botanical Experiments, helped to generate interest in the possible economic exploitation of the frozen continent (Albert 1901; Pinochet de la Barra 1976, 82–85). Acting through the Ministry of Industry, the government adopted a policy which equally prioritized economic exploitation, administrative prerogatives, and the promotion of meteorological, zoological, and botanical research (see Chilean Decrees nos. 2074 and 2305 of October 1903 and Decree no. 3310 of December 1902).

Chilean interest in Antarctica in the early twentieth century

The Swede Otto Nordenskjöld accepted Chilean assistance and, while in the country, he and the president of the Chilean Scientific Society, Federico Puga Borne, tentatively planned a joint expedition to the Southern Shetlands (Barrera 1983). When Nordenskjöld continued to Ushaia, Argentina, in 1902, he discussed the possible expedition further with Ismael Gajardo (1905a; 1905b), a Chilean naval officer who had gathered much related information and published a detailed account in the widely read *Revista de Marina*. The following year the government sent Lieutenant Alberto Chandler to assist the rescue of Nordenskjöld, as well as to collect data related to the area's potential resources (Matte 1903; Escudero 1953; Braun 1974, 146; Pomar 1903). However, the joint Nordenskjöld and Borne expedition failed to transpire. In 1906, the Chilean government renewed its interest in the far south, this time seeking to establish permanent occupation by constructing a meteorological and magnetic station. This new plan coincided with the influence of the Scientific Society of Chile, specific military officers, diplomatic officials, and Luis Aldunate who served as an assistant to the preparatory meeting for the First Polar Year (Vila 1947, 15; Huneeus 1948, 43; Berguño 1999, 5). Aware that this endeavor would most certainly be of interest to Argentina, the government initiated a series of negotiations to delineate Antarctic boundaries between the two nations (Berguño 1999, 11).

Chile's Antarctic consciousness was further promoted by Minister of Education Julio Montebruno (1909), who attended the Ninth Geneva Congress on Geography in 1908, and sent numerous reports to the government and published related material in the *Anales* of Universidad de Chile. Monterbruno's work accompanied by other newly acquired geographical knowledge began to appear in texts and atlases used in the nation's secondary schools, including *Compendio de Geografía Descriptiva* by Manuel Salas Lavaqui; *Compendio de Geografía General* by Justo Parrilla; *Nueva Geografía Universal* by Vivien de Saint-Martin; *Nueva Geografía Universal Arreglada para los Colegios Americanos* by José Manuel Royo; *Geografía para la Enseñanza Secundaria* by Gonzalo Cruz; *Allgemeiner Handatlas* by Richard Andrés; *Geografía Universal* by Emilio de Medrano; and *Atlas de Geografía Universal* by Juan Roma.

Once again, the government's plan for an expedition had to be set aside, however, namely because of financial considerations. Yet Chile maintained activity in the sub-Antarctic waters during the First World War, most notably with the rescue of Sir Ernest Shackleton's expedition (González 1966; 'El Piloto' 1941; 'El Salvamento' 1950; *Honorable Cámara de Diputados de Chile*, 2nd Session, 26 November 1946).

The interest in Antarctica's territorial and resource-related potential grew in the 1920s. By this time, the Chilean National Society of History and Geography had become an important center of knowledge, publishing a series of maps prior to 1911, and thereafter Chilean publications frequently praised the success and heroism of the Chileans active in the sub-Antarctic waters (Almeida 1956; Edwards 1911; González 1966). In September 1916, the Society welcomed Shackleton, and awarded a medal of honor to pilot Luis Pardo Villalón, who had rescued him ('Manifestación' 1916; 'Homenaje' 1956). In this era Lieutenant Ramón Cañas Montalva began to closely follow global developments in Antarctica while actively championing Chilean rights. Cañas soon became an influential advisor (Berguño 2003, 43), as he would remain throughout the 1957–58 IGY. The Military Geographic Institute, created in 1922 and subsequently an active member of the International Geographical Union, also began to foster the nation's knowledge of and interest in the Antarctic.

The Second IPY (1932–33), sponsored by the International Meteorological Organization (IMO), aroused widespread enthusiasm despite the global depression ('Temporary Commission' 1950). The Chilean Meteorological Office assisted the event, establishing a preparatory commission ('Chile' 1932). Chilean Supreme Decree no. 2031, issued 30 November 1931, created the preparatory commission. The following March sub-commissions were established devoted to meteorology, solar radiation, magnetism, seismology, atmospheric electricity, auroras, communications, the upper atmosphere, and cooperation and publications. However, for economic and bureaucratic reasons, the compilation and analysis of the data had to be postponed until late 1947, at which time these objectives were completed with a grant from the Rockefeller Foundation, based in the United States, and soon thereafter records from the stations in the Magallanes region, and elsewhere, were published ('Temporary Commission' 1950).

The Second IPY was motivated, to some extent, by the 1933 Antarctic expedition of Richard Evelyn Byrd, which had received the full support of President Franklin D. Roosevelt (1933). At the time Byrd was credited with having 'single handedly' renewed US interest in the frozen continent after approximately ninety years of neglect (Siple 1955). Chilean Lieutenant Patrick Wiech Mann accompanied Byrd on his 1928 expedition, and sustained US interest in the south polar region would soon prompt the Chilean government to take more decisive moves (Rodríguez 1978–79; 'Expedición' 1955). In 1939, Washington reached the conclusion which the Chilean government had reached over three decades earlier, but thereafter proved unable to act upon: that the minerals believed to lie under the ice cap would be extremely valuable. Therefore sending an expedition would also help to reinforce the government's basis for a sovereignty claim. Equally important was that the Germans had recently mounted a large expedition and appeared likely to establish their own bases, as well as to formalize their rights on the continent. During the 1939–1941 Byrd expedition, accompanied by two Chilean lieutenants, the Chilean government officially declared its sovereignty over the sector from 53° to 90° West (Escudero 1984; Tromben 1997a, 2–3). Upon his return, Byrd was received by Chilean officials and the Society of History and Geography published a long interview with him regarding the geological contiguity between the Andes Mountains and the Antarctic Peninsular region as well as possible economic exploitation of the continent (Barrera 1940).

Chilean research in Antarctica

Throughout the 1940s, Chile's interest in the Antarctic centered on domestic national maritime and territorial integrity issues as well the continent's potential resources. In 1942, the Chilean government created an Antarctic Commission to 'encourage studies, research, surveys, and exploration which might be thought suitable for the fuller exploitation of the said territory' ('Chilean Decree' 1950; 'Se Fijaron' 1940), and sent observers with the 1943 Argentinean expeditions (*Honorable Senado de Chile*, 23rd Session, 13 January 1954; *Honorable Cámara de Diputados de Chile*, 2nd Session, 26 November 1946; Cordovez 1946). Shortly following the Second World War, possible conflicts of interest between the various nations with interests in the Antarctic grew more pronounced. The newly elected Chilean president, Gabriel González Videla, a member of the Radical Party, initiated an educational program emphasizing the territorial integrity of Chile, both on the mainland and in the Antarctic. In spite of the economic turmoil confronting his administration, he devoted time and resources to consolidating the nation's presence on the frozen terrain. According to Oscar Pinochet de la Barra, former director of the Chilean Antarctic Institute, the president's decision to construct a permanent base was motivated by Admiral Richard E. Byrd's announcement, in November 1946, that the US government refused to recognize the claims of other nations (Pinochet de la Barra 1997). For this reason scientists from the University of Chile and the University of Concepción quickly convened to finalize plans to establish a meteorological-magnetic laboratory and possibly one devoted to seismology (Tromben 1997b). The Chilean public and its elected officials were aware of the United States' interest in the continent's mineral resources as well as its non-recognition of Chilean national rights. Ramón Cañas Montalva, then commander-in-chief of the army, increased his participation in various scientific institutions while gaining still greater influence among civilian policymakers (Barrera 1977; Pinochet de la Barra 1996). As Foreign Minister Raúl Julliet remarked in January 1947, the forthcoming expedition sought to expand the smaller republic's 'scientific, strategic, and economic knowledge of the region' (*Honorable Senado de Chile*, 16th Session, 21 January 1947).

While many nations with interests in the Antarctica and sub-Antarctic waters proclaimed that science was their primary motivation, journalists ventured that to various degrees they were all the involved in a 'uranium race' (Moore 2001, 725). Attempting to dispel this perception in relation to the United States, Byrd placed greater emphasis on science in his interviews with the press. According to sources, he viewed the early postwar period as the ideal time 'time to act while we have trained manpower and excess equipment'. He persuaded the administration of President Harry S. Truman of the desirability of converting the Antarctic into a 'scientific laboratory' as well as a testing ground for military equipment and transportation (Byrd 1948a), a dual objective in which top US officials saw no contradiction (Joint Chiefs of Staff 1950). Byrd proposed leading a new expedition to reaffirm the United States' position as 'a leading nation of power and scientific research', which thereby had the responsibility to lead the effort to exploit the planet's untapped resources (Byrd 1945). Aware that Chile had already established a permanent presence in the Antarctic, he urged the US government to devise some means of seizing its control, preferably by 'fair and peaceful means' and without incurring 'accusations of imperialism' (Byrd 1948b).

The problem confronting the United States was that the territorial dispute over the Antarctic seemed capable of dividing its Cold War allies, specifically the Southern Cone nations and Britain, whose three set of claims overlapped in the peninsular region. While the Russians, like the 'Americans' (as citizens of the United States often refer to themselves), had not announced a claim, they were whaling in Antarctic waters and, also like the 'Americans', were fully capable of reversing their non-claimant position. *The New York Times* noted that what the Southern Cone nations, whose citizens considered themselves equally 'American', referred to as the 'American Antarctic' did not refer to the regions explored by the United States but rather to the sector in which their own interests lied ('Antarctic Region' 1946). Meanwhile the United States dispatched two more expeditions – one commanded by Byrd and the other by Finn Ronne. The Department of State repeated that the government did not recognize the claims of any other country and reserved all rights in those and all other areas ('Antarctic Rights' 1946; 'United States' 1949). Surprisingly, unlike most official statements, this particular statement contained no reference to the ostensibly scientific motivation of US Antarctic policy.

In early 1947, the Chilean government sponsored an expedition, the primary objective of which was 'to report on scientific, naval, and military matters, and to establish and occupy a meteorological and magnetic station' ('Chilean Antarctic' 1953; Tromben 1997b). The expedition was placed under the command of Captain Federico Guesalaga Toro, director of the Hydrographic Department of the Chilean Navy. Its members included Major Pablo Ihl Clericus, geodetic and topographical surveyor and Próspero Madrid, a topographical surveyor – both of whom were affiliated with the Military Geography Institute – as well as fourteen civilian scientists and technical experts ('Chilean Antarctic' 1953; 'Los Que' 1997). The press noted that these individuals appeared to be extremely capable and it was hoped that their findings would be far more useful than those associated with a traditional laboratory setting. One article ventured that that the mineral-related discoveries might help to determine the shape of 'our industrial future' ('Llegó Hoy' 1947).

Responding to indications that, as in past centuries, the larger powers continued to place a highly military emphasis on their 'scientific' activities, as well as seeking to reinforce their presence in the Antarctic, in October 1948, the Chilean government established an Antarctic Section within the General Staff of the Armed Forces, vesting it with control over both military and scientific issues (Chilean Decree no. 1168 1950). Furthermore, the government recommended negotiations with Argentina to delineate a mutually satisfactory boundary in the Antarctic sector from 53° to 74° West in which their territorial claims overlapped (Chilean Decree no. 548 1950). President González Videla also ordered that all school materials related to the Antarctic be promptly updated (*Honorable Senado de Chile*, 16th Session, 21 January 1947).

Despite growing public awareness of the Antarctic and support for preserving the nation's south polar heritage, a relatively limited number of people – individual military officials, academics, diplomatic personnel, and congressional representatives – composed what could be referred to as the nation's Antarctic elite. Its members shared what C. Wright Mills (1956, 277, 288) once referred to 'a coincidence of interests', in this case linked to a distinctly non-clerical sector of the society. This phenomenon contained elements of what Paul W. Drake (1978, 85) has referred to as modern 'corporativism and functionalism'. Though many Chilean scientists sought to expand their nation's presence in Antarctic, Ramón Cañas Montalva (1945) was foremost among them and his concept of an 'Austral-Antarctic Zone' exerted widespread influence. Meanwhile González Videla possessed and utilized the ability to convert the interest of his nation's Antarctic elite into an anti-imperialist crusade, directed primarily against the British, which was supported by all political sectors ('El Presidente' 1948).

The president's declarations during and after his trip to the Chilean Antarctica in 1948 were enthusiastically received by Latin Americans, nearly all of whom resented the British presence in the Western Hemisphere and concurred, at least rhetorically, that Chilean territorial rights extended to the South Pole. It was also evident that some form of international agreement would be needed to avoid any dangerous incidents related to the territorial dispute ('Territorial Claims' 1949; 'Antarctic Claims' 1949). After a proposal to forge a condominium arrangement with the seven claimant nations – Argentina, Australia, Britain, Chile, France, New Zealand, and Norway – failed to gain support, the United States continued to pursue the 'acquisition of scientific data' as a means of reinforcing its military operations in the Antarctic and its desire to evaluate its 'inventories of natural resources' (Early 1949). Aware that the question of sovereignty might derail further attempts to achieve an international agreement, Washington cancelled its plans for another Antarctic expedition, lest it appear to confirm that its designs on the frozen continent were imperialistic in nature (Byrd 1949).

In the latter part of the 1940s, Chileans grew alarmed by possible conflicts of interest with the United States. Mario Rodríguez, the Chilean attaché in Washington, clarified that this related to indications that Washington had resolved to send 'scientific and technical expeditions' to as much as the territory south of 60°S as possible and to withhold all related information to advance its unilateral interest (Department of State 1977, 905, 908). While many Chileans believed that the leader of the so-called 'free world' was primarily motivated by a desire to exclude the Soviet Union and as many sympathized with that objective, they feared that their own interests might be subordinated to Cold War rivalry. This was exacerbated by the United States' initial reluctance to offer Chile modern naval vessels for use in the sub-Antarctic waters and elsewhere at competitive prices (US Embassy 1950).

Throughout the following decade, Chilean scientists in the Antarctic continued their routine work and opened two more bases or 'meteorological stations' ('Antarctic Claims' 1952). Many of the scientists must have regarded these activities as intrinsically valuable, yet most Chileans somewhat mistakenly regarded them in political terms. As Miguel Cruchaga Tocornal explained to the congress, the prerequisite for acquiring rights over the polar regions – according to some interpretations of international law – was occupation (*Honorable Senado de Chile*, 18th Session, 22 January 1947). He clarified that, while the activities currently underway could be viewed as establishing that prerequisite, they were being 'carried out within the limits of the jurisdiction of the Republic and in the exercise of our sovereignty'. The Chilean scientific program, unlike that of the United States, did not seek to validate the rights officially declared in 1940 and taken for granted since colonial times; rather the government viewed it as an extension thereof ('Antarctic Claims' 1952).

Chilean claim to Antarctica

A large part of Chilean scientific activities continued to seek to establish the geographical contiguity between the Andes Mountains and the Antarctic Peninsula. This notion gained currency also because it was thought to be helpful in relation to boundary negotiations with Argentina (*Honorable Cámara de Diputados*, 10th Session, 18 May 1955; Bruggen 1953.) It had been accepted by members of the Chilean Antarctic elite as early as the 1930s (Pinochet de la Barra 1944, 25–42; Ihl 1953; Duran 1982; Romero 1985), and later by Cambridge geologists Priestley and Tiller as well as by US Admiral Richard E. Byrd. Moreover, it was formally presented by Chilean scientists at the Tenth Assembly of the Geodesic and Geophysics International Union, held in Rome (Kosack 1955; Cañas 1956).

In the early 1950s, Chile and Argentina opened new bases in the Antarctic archipelago to dispel British hopes that they might retreat from their territorial claims. Argentinean President Juan Domingo Perón forcefully stated his resolve to defend the smaller nations' rights, and military personnel acted accordingly, preventing a party of British geologists from landing on the disputed territory ('Argentine Fire' 1952), an action which Perón and many Chileans viewed as contrary to the Rio Treaty of 1947. The Rio Treaty demarcated a defensive perimeter around the Western Hemisphere that the Southern Cone nations believed extended to the South Pole. Instead the United States dwelled only on its objective of seeking to exclude the USSR, which they believed might be used as a position in the Antarctic to facilitate an infiltration into the Western Hemisphere (see Miller 1950; Miller 1959). For this reason, Washington remained mute when the British retaliated by dismantling the small Chilean and Argentine outposts on Deception Island and resisted pressure to take the incident before the Organization of American States (Moore 2003, 173–75). Considering the dispute between its Cold War allies, and US incertitude about how to consolidate its Antarctic interests, the Eisenhower Administration chose to modify the thrust of its Antarctic policy, now stressing the role of science and its contribution to world peace.

In mid-1953, the Joint Chiefs of Staff (1954) was concerned that the increasing activities of other nations in the Antarctic might serve to undermine the basis for the United States' unannounced rights. The following month President Eisenhower announced that, after many years of inactivity, the United States had begun to plan a new expedition (Transcript 1954). Later that year he consulted at length with Byrd whose opinions influenced the decisions of top officials. It was decided to dispatch the preliminary expedition as soon as possible and then to proceed with a large-scale expedition the following season. The US press recognized the importance of this decision and started referring to the Antarctica as 'the last frontier' (1954).

Chileans received limited information of the preliminary expedition shortly before it departed. Journalists noted that Washington defined its objectives in terms of logistical issues and conducting geophysical experiments, objectives which, it claimed, did not entail establishing a new base ('Torneo Mundial' 1954). No further details were known, not even that the expeditions planned to be active in the Chilean Antarctic territory or, as the Australian press had speculated, that its agenda might include conducting nuclear tests (Foreign Office 1954; Department of State 1956c). Chileans journalists even failed to discern that the expedition was part of preparations for the IGY; in their perspective, it was simply another US expedition.

After the preliminary US expedition was underway, the Chilean press followed it closely and made the connection with the IGY. According to Walter Sullivan (1955), science editor of *The New York Times*, the preliminary expedition was placing greater emphasis on scientific research than Operation High Jump, the US Navy's massive 1946–47 expedition, even though only eleven of the 267 crew members were scientists. In March 1955, Chileans discovered that later that year there was to be another US expedition, Operation Deep Freeze I, led by Byrd, to determine the most useful locations to establish two more US bases – a number which finally rose to six, their ostensible purpose being to collect data for the IGY. As much as some Chileans wished to believe that their 'good neighbor' was philanthropically motivated, the expedition's military nature could not be denied, for it included five war vessels, seven aircraft, and approximately 1,400 crew members, only a fraction of whom were scientists ('Estados Unidos' 1955a; 'El Gigantesco' 1955).

United Press dispatches dwelled on scientific rather than military issues, and Chileans gathered that their nation had little to contribute in either area. The coverage implied or stated that the IGY was giving the superpowers the opportunity to compete for prestige ('Estados Unidos' 1955b; 'La Exploración' 1955; 'Gran Bretaña' 1955). However, it had been agreed by all IGY parties that the bases constructed before and during the event would not give the participating nations the basis to claim territorial rights. It was equally clear that the bases were likely to remain after the event possibly diminishing previous understandings, thereby dragging the continent into the Cold War ('En 1958' 1955.) For example, Australia, a close ally of the United States, permitted the Soviets to be active in the sector over which it declared territorial rights. While Chileans preferred this to Soviet activity in the peninsular region, they remained convinced that the rhetoric of science was being used to disguise political objectives.

In order to placate these fears, Byrd announced that Chile would be the first nation he would visit when returning from Deep Freeze I. Despite his long-standing popularity with Chilean officials and the public at large, journalists continued to speculate that the United States was preparing to reverse its policy in favor of an exploration-based territorial claim ('Expedición' 1955). Such pessimism appeared to be confirmed by the fact that before returning home US explorers raised their flag over the South Pole, the point at which most of the claimant nations' territorial ambitions converged ('Estados Unidos' 1955c). Chileans grasped that the United States was pursuing an agenda that easily might be used to advance its unilateral interests. While its primary objective of appeared to be preventing Soviet advances, Chileans dreaded that their own rights might fall prey to superpower rivalry.

In late 1954, President Eisenhower encouraged Chile to participate in the IGY (Sullivan 1954). The smaller American republic had little choice but to accept, for it required US economic assistance and was attempting to follow the advice of US consultants to increase the efficiency of its copper industry, the most essential source of government revenue, which was disproportionately controlled by US companies (Urzúa 1968, 94). President Carlos Ibáñez del Campo had devoted most of his two years in office to efforts to curtail the rate of inflation and cost-of-living increases, and had found it necessary to retreat from his 'anti-American' campaign pledges. Several other factors added to his unpopularity. For example his reduction of the government's bureaucracy and his desire to forge a closer economic and political alliance with Argentina; a nation that most citizens despised as much or more than the United States (Olavarría 1965, 153; Würth 1958, 343, 354). Compared to these issues, Chile's participation in the IGY was not crucial.

This is not to say that Ibáñez had neglected the frozen continent. He had successfully urged the passage of legislation reorganizing the administrative status of the Chilean Antarctic and reinforcing the traditional perception of geographical contiguity between the nation's southernmost continental province, Magallanes, and the Antarctica (*Honorable Senado de Chile*, 52nd Session, 17 May 1955). Likewise, an issue of even greater importance for the public was the nature of British activities. Though there had been no further hostile encounters since the Deception Island episode in 1953, Chileans viewed the British, rather than the Soviets, as posing the most serious threat to their Antarctic rights. They believed that, like the North Americans, the British were using purportedly scientific objectives to attempt to solidify their rights in the peninsular region and its vicinity. Of particular concern were reports that the next British expedition sought to 'determine the [continent's] scientific and mineral potential', and might do so by using 'a series of explosive charges from one side of Antarctica to the other in an attempt to plumb the depths of the polar icecap' ('British Antarctic' 1955; 'Gran Bretaña' 1955).

Concerned with domestic problems, the Chilean public did not dwell on the IGY, though many remained dubious of US and British claims to have purely scientific interests in Antarctica. What seemed evident to even casual observers was that the Anglo-Americans sought to evaluate the potential mineral wealth while closely monitoring Soviet activities and devising plans to meet any communist challenge at the bottom of the world ('La Guerra Fría' 1956). These perceptions were both correct and understandable since, by this time, the superpowers were engaged in a full-fledged nuclear arms race. Chile sought to maintain as much distance as possible from the Untied States while not having to turn to the Soviet bloc for economic assistance.

The conditions of US economic assistance were not popular, yet neither were they completely unrealistic. For example, the United States did not insist that Chile send troops to the Korean War, an alternative which no doubt would have ignited virulent protests from all sectors of Chilean society ('Antarctic Riches' 1954). After that conflict, the United States began to assume a more conciliatory stance toward the USSR (Sullivan 1954). Though in years past, the United States would have sought to exclude the USSR from any involvement in the far south, the 'fall of China', stalemate in Korea, and Soviet development of thermonuclear weapons had encouraged officials to be more flexible.

The United States recognized that Antarctica was one of the few regions of the planet that did not clearly fall into any single power's sphere. While further research pertaining to its natural resources might prove beneficial from a unilateral prospect, it might be counterproductive to risk transforming the continent into a new Cold War front (Department of State 1958). In late 1955, Secretary of State John Foster Dulles recognized that his nation's non-claimancy, non-recognition policy had in some way encouraged Soviet involvement, for the USSR had adopted the same policy and as such could claim equal standing with the United States. Due to a translation of an article from *The New York Herald Tribune* which maintained that the United States' 'preponderance in research and explorations' had given it 'a privileged position in the Antarctic' ('Problema' 1956). Chileans' remained concerned that Washington still might reverse its position and announce the territorial claim. Soon thereafter, in 1957, the Chilean public also learned that Senator Thor Tollefson, with the full support of Byrd and Ronne, called for making a national claim since the United States 'has dispatched more expeditions ... than all the other nations together' ('Reclamará' 1957).

This position found some support within the Eisenhower Administration: in April 1957, the Joint Chiefs of Staff (1957) recommended making a US claim or claims over the largest possible area, including those already claimed by other nations (Joint Chiefs of Staff 1957; Department of State 1955a). However, officials had to recognize that such a course would inevitably antagonize the Soviet Union as well as the seven claimant nations, especially Chile and Argentina. It was decided that the best alternative was to place greater emphasis on science, as this would serve to justify further US expeditions while generating less controversy, and it would also enhance US scientific prestige, an objective which officials thought would be of 'considerable propaganda value in the Cold War' (Department of State 1955b).

US hesitation to advance a territorial claim was perhaps most related to concerns about the spread of communism in Latin America. A US claim would have appeared to confirm Soviet propaganda that Washington sought to militarily and economically subject the continent to its imperial designs. As senior officials recognized, the small scale of US economic assistance to Latin America had been 'inadequate to meet the intensified Soviet challenge and new Soviet tactic in Latin America', precipitating the rise of nationalistic leaders who often advocated closer ties to the Soviet Union (Operations Coordinating Board 1956). US officials believed that the heated protests surrounding Vice President Richard Nixon's visit to Latin America attested to the existence of a 'communist menace' in Latin America (Operations Coordinating Board 1958). One means of containing this was to withhold an Antarctic claim.

Chile and the IGY

In addition to sending official representatives to the IGY preparatory meetings, Ibáñez del Campo established a national commission, headed by Ramón Cañas Montalva, which coordinated the Antarctic-related work of universities, the armed services, and scientific institutes. Cañas had continued publishing articles emphasizing the importance of Antarctica and followed the progress of scientific meetings held in Europe. Though he himself was not a scientist, he served as a link between the scientists, military personnel, and politicians. Unfortunately, his best efforts coupled with the support of Ibáñez were insufficient to ensure that there would be substantial funds for Chile to participate in the IGY. The visit of the US icebreaker *Edisto* in late March 1956 helped Cañas and Ibañez to demonstrate the importance of going to Antarctica ('Visitas' 1956; 'El Edisto' 1956). After a meeting with his cabinet, the president created an Antarctic museum as well as an Antarctic Department within the Foreign Ministry ('Consejo' 1956).

Predictably enough, congress approved an extremely modest budget for the IGY, and Ibáñez continued his personal initiative to secure funds from every other possible source for the sake of acquiring additional scientific equipment and the materials to build a new Antarctic base (*Honorable Cámera de Deputados*, 12th session, 18 April 1956). The visit of another US icebreaker, *Glacier*, again encouraged the press to turn its attention southward and venture that an upcoming Chilean expedition would help to reaffirm, before the entire world, the undeniable nature of Chilean rights (Escobar 1956; 'Año Geofísico' 1957a). While some journalists encouraged congress to expand its funding of the national IGY program, congress stood firm, even refusing to release the approved sum until the end of the year, by which time all of the vessels composing the Chilean 'expedition' had already sailed ('Año Geofísico' 1956; 'Año Geofísico' 1957b).

This slowness and meagerness of the congressional appropriation was related both to the unpopularity of Ibáñez and limited knowledge and interest pertaining to science in general and particularly as related to the Antarctic. Congressional representatives spent long sessions extolling their nation's inviolable historical rights, yet they very rarely mentioned the intrinsic or political value of increasing their research effort (*Honorable Cámera de Deputados*, 2nd Session, 20 October 1953; 13th session, 16 June 1955; *Honorable Senado de Chile*, 26th Session, 21 January 1947). In 1953, for example there had been consideration of using Antarctic science to bolster the nation's position on the continent, hopefully leading to international recognition of its rights (*Honorable Cámera de Deputados*, 36th session, 22 December 1953). However, curiously this did not reemerge on the eve of the IGY. Neither did the hopes previously expressed by journalists that the region might serve as a valuable source of mineral resources, including uranium – a position formerly championed by Ignacio Palma Vicuña, leader of the later influential Christian Democracy Party (*Honorable Cámera de Deputados*, 13th session, 16 June 1955; 26th session, 12 July 1955).

Nonetheless, a general feeling persisted that Chile's scientific activities, however limited and poorly funded, might help the nation gain knowledge of the larger powers' endeavors which were conducted with virtually 'unlimited human and economic resources' (*Honorable Senado de Chile*, 32nd session, 22 March 1961; 33rd session, 4 April 1961) In this manner, it was hoped, Chile might reap benefits from the IGY without having to contribute on a grand scale. During 1956 the press carried numerous reports on developments which greatly impressed readers, aware as they were of their essentially peripheral involvement. In the region known as Little America, the United States was building yet another base to be equipped 'with all possible comforts and the latest scientific innovations' ('Listo el Lugar' 1956), while the Soviets were constructing a long highway to facilitate their own activities, such as an expedition deep into the unknown interior and the launching of research satellites ('Los Rusos' 1956). Chileans vicariously experienced the thrill of such events by offering their ports to the US Navy and in return receiving an invitation to send an observer (Department of State 1956a).

Chilean Antarctic scientists continued to work on a limited budget and with little public awareness of their work. Beginning in 1953, some Chilean scientists had received governmental assistance to attend scientific conferences in Europe. At one of the most important, the 1955 meeting in Paris, Chilean researchers encouraged a resolution that future Antarctic meetings would be relegated to scientific and not political issues ('Chile Dio' 1956). Unfortunately, the IGY was less beneficial than anticipated in raising the Chilean profile, either nationally or internationally. One of the ways they sought to gain public attention and secure more government funding was by linking the nation's traditional boundary dispute to the theory of geographical contiguity between the Andes Mountains and the Antarctic Peninsula (see Muñoz 1948; Fairbridge 1955; Ihl 1953; Kosack 1955; Cañas 1956). The validity of this position, as explained by Humberto Fuenzalida and Pablo Ihl, gained recognition at various international scientific conferences (*Honorable Cámera de Deputados*, 10th session, 18 May 1955).

The Chilean government's agreement to participate in the preparations for the IGY, as well the 18-month event itself, entailed a number of responsibilities that involved the armed services. The corresponding officials had neither been previously consulted nor given addition funds to build additional bases or carry out scientific research, for which their personnel were insufficiently prepared. The navy's Physical Oceanography Division was only able to complete its surveys with assistance from the Scripps Institute of Oceanography, an organization based in the United States (see 'La Oceanografía' 1962). In the 1954–55 Antarctic season the navy constructed a meteorological station on Deception Island. The following season this was augmented by an airfield and pier ('Commandante' 1956; 'En la Antártica' 1956), and collaboration was maintained with the Universidad de Chile's Institute of Marine Biology (Sievers 1968).

In March 1956 the navy installed a station at Punta Arenas in order to facilitate the research of all scientists contributing to the IGY ('Importante Aparato' 1956). This helped to counteract the concern aroused by its activities on Peter I Island.

Although they had been carried out for the same purpose, the island lay within the Antarctic sector claimed by Norway, and thus had evoked diplomatic protests ('En Dos Meses' 1956; 'No se Justifican' 1956). This was also exacerbated by reports that Chile was seeking to purchase an icebreaker from Germany ('Se Ordena' 1956). The ice-breaker, it was feared, might be used to enforce Chilean rights over the territory it claimed as its own (the negotiations for this transaction failed to reach fruition. 'Chile ha Ordenado' 1956.

During the 1956–57 season, the navy transported 41 Chilean scientists to the Antarctic for the winter ('El Rancagua' 1956). Though the national press followed the navy's activities more closely than those of the air force and army, the latter branches of the armed services remained active, for example, installing a seismographic station on *Base O'Higgins*. In January 1957, the army also proceeded with the construction of the Bases Gemelos, which were located farther south, and thus were of even greater utility for both the nation and the IGY ('Dos Nuevas Bases' 1957). Meanwhile the air force had initiated a series of flights between Punta Arenas and Decision Island ('Quedó Completado' 1955).

Despite the protests of some members of the foreign affairs ministry who failed to appreciate the value of their nation's participation in the IGY, in January 1957 the government announced that Chile would construct yet another base (Franulic 1956). This one was designed to be used exclusively for scientific research and managed by civilians affiliated with Universidad de Chile, Universidad Católica de Chile, and Universidad de Concepción ('Dos Nuevas Bases' 1957). Upon its inauguration that March, Foreign Affairs Minister Osvaldo Saint-Marie referred to the nation's scientifically based rights deriving from the geographical contiguity between the Andes Mountain and the Antarctic Peninsula; the ministry also emphasized that international cooperation would in no way alter 'the juridical status of Antarctica' – an idea which some other nations apparently hoped to change in their favor due to their participation in the IGY ('Chile Inagura' 1957).

Ambassador Juan B. Rossetti, a well-known defender of Tierra de O'Higgins, as the Chilean Antarctic was known, presented the nation's modest scientific contributions at the Paris Antarctic meeting in June 1957. His presentation dwelled on the gravimetric, meteorological, and aurora-related studies being conducted in Punta Arenas and on various bases in the Antarctic archipelago. One of these bases – Base Risopatrón – was only partially reconstructed following a catastrophic fire ('Chile prestará' 1957). The public remained more concerned with the presence of eleven Argentine camps on Chilean territory ('11 Campamentos' 1957) while the United States proceeded with high-technology observations at Antofagasta, Quintero, and Puerto Montt.

The national press did not closely follow the progress of the thirteen groups of Chilean scientists stationed at five bases, the most noteworthy of which included Palmenio Yánez, a marine biologist affiliated with Universidad de Chile. Yanez had been engaged in Antarctic research since 1947 and participated in the IGY Preparatory Commission ('Los Que' 1997). Other biologists included Guillermo Mann and Alberto Andrade, who studied sea-water circulation, as well as Nibaldo

Bahamondes, who researched seals (Romero 1986, 43). In addition, there was Humberto Barrera, a member of the Chilean Scientific Society affiliated with Universidad de Chile, and glaciologist concerned with change in global weather patterns ('Trabajos' 1962); geologist Humberto Fuenzalida, also affiliated with Universidad de Chile whose interests were the Shetland Islands in particular, a field which would later be pursued by Francisco Hervé (Romero 1986, 43).

The scientific activities of the armed services increased substantially in the 1957–58 season. Due to a limited budget, they concentrated their activities at Base O'Higgins. The personnel believed that work in the areas of meteorology, geomagnetism, and oceanography, for example, would help to reinforce the nation's sovereignty as well as demonstrating to the world that the Chilean armed services were willing, if not able, to prevent other nations from encroaching on the nation's territory, or at least a few small bases located therein (Reyes 1957).

While this nationalist perspective was prevalent among most Chileans working in the Antarctic, civilians at Universidad de Concepción sought and thankfully acquired US assistance to conduct a series of ionospheric surveys and marine biology experiments. These and other individuals remained cautious not to make declarations which might be interpreted as welcoming the 'internationalization' of Chilean territory, for indeed this was one of issues of the greatest concern to the public (see 'Mar Territorial' 1954).

In February 1958, the press reported that the British Ambassador in Santiago had presented the government with a proposal to establish the joint administration of the disputed peninsular region. Alberto Sepúlveda, the new foreign affairs minister, firmly rejected its call for dismantling Chilean military bases since, aside from affronting the nation's sovereignty, that course would also hinder its scientific progress and cooperation with the other IGY participants. It was well established at this point that the armed services had assumed distinctly scientific responsibilities ('Internacionalización' 1958). Sepúlveda's rejection gained front-page headlines while journalists returned to the nation's long-held position – which assumed that the hemispheric defense provisions of the Rio Treaty of 1947 extended to the Chilean Antarctic; a position which neither the US or Britain accepted. In relation to the British proposal's call for 'demilitarization', Sepúlveda insisted that the current informal agreement between Chile, Argentina, and Britain to refrain from hostile naval displays was sufficient ('Chile Recharzará' 1958).

That May, three months later, the United States circulated a more elaborate proposal to the twelve nations participating in the Antarctic component of the IGY. It incorporated aspects of the British proposal but not the requirement for military bases to be dismantled. For this reason and the evident need to reach some form of agreement to forestall possible hostilities, the US proposal served as an acceptable basis for the commencement of preliminary negotiations in Washington. While Chile did participate in these negotiations and the Antarctic Conference held the following December in Washington, its representatives did not assume a dominant role. Neither was Chile given a convincing reason to do so. The treaty under discussion permitted nations to leave their territorial claims in place while continuing their peaceful scientific cooperation – consistent with the Chilean Escudero Plan.

The treaty's final inclusion of a nuclear test ban and the prohibition of the disposal of nuclear waste helped to dispel Chilean concerns that the southern continent might yet be drawn into the Cold War. However, some politicians opposed the treaty's ratification since, for example, it permitted the use of nuclear reactors, which held the very real potential to contaminate to continent ('Chile' 1959). Another point of unease related to the freedom of scientific research, which, as Argentina recognized (Mora 1961), severely undermined the traditional notion of sovereignty. Of equal, if not greater concern to Chilean scientists was the perpetual lack of resources.

The Chilean government signed and thereafter ratified the treaty, for doing otherwise would have would have excluded it from one of the most durable agreements in history. Unfortunately, few Chileans had recognized the importance of their nation's scientific activities during the IGY. Unlike Ramón Cañas Montalva, the general public remained unconvinced and failed to grasp the possible benefit of science as a means defend Chilean territory and waters. Cañas (1956–57) regretted the disinterest – and occasional contempt – of some government agencies toward these issues (see Franulic 1956). Most scientists were equally chagrined by the minimal amount of human and material resources, which they received, which in turn hindered their contribution to the IGY. Nonetheless, they were able to gain valuable experience and establish mutually beneficial relationships with foreign and predominantly US scientists and organizations. The field of oceanography, in particular, benefited most clearly as a result (Sievers 1968). Immediately after the IGY, the air force transferred its base and equipment to the Universidad de Chile and in 1963, all bases and equipment were placed under jurisdiction of the newly established Instituto Antártico Chileno (González-Ferrán 1991).

Conclusion

From the earliest days of independence through participation in the IGY, Chile's Antarctic mentality remained fixated on national sovereignty. Despite the consistency of this theme, few citizens during the 1950s felt any need to invest substantially in Antarctic science or even to expand the nation's military presence. Indeed the geological, legal, and historical basis for Chile's Antarctic rights was not contingent upon these objectives. However, it remains possible that a more robust scientific program might have enhanced Chile's prestige and indirectly generated sympathy for its rhetorical devotion to the far south. While there is much evidence to suggest that the British perceived Chilean science as 'third rate' reflecting a political or cultural bias, it certainly was not 'first rate.' Perhaps some consolation can be taken in overall British perceptions of the US Antarctic program as 'mediocre at best' (see Department of State 1956b). If the same or worse was true of the Chilean program at least it could be attributed to a scarcity of resources.

Chile's participation in the IGY did not greatly diminish its traditional leeriness of science. This leeriness related to the non-profitability of science and Antarctic science in particular, as well as to the fear to the unknown. To this day Chileans find it difficult to appreciate the corollary – or possible corollary – between scientific activities and the defense of their Antarctic rights. They still believe that the nation's permanent occupation provides one of the strongest bases for its rights. What they do not fully appreciate is that by shifting emphasis to international scientific cooperation, the larger powers gained for themselves the perfect means by which to justify their own permanent occupation and determine which areas eventually might be suitable for exploitation.

Thousands of scientists from dozens nations are participating in the current IPY and it is hoped that their discoveries will reveal means of sustaining the global environment ('International' 2007). Though 'change is ubiquitous in Earth's history', it is not always positive. At present the ice shelves have begun to recede while many glaciers are disappearing altogether. *Science* magazine entertains hope that new technological and conceptual innovations will help to mitigate such developments (Albert 2004). There is perhaps even a greater opportunity for international scientific cooperation today than during the IGY when the Cold War overshadowed all else ('The Ends' 2007). Though Chile's contribution to the IPY is modest, its scientists have gained the opportunity to demonstrate their aptitude before the rest of the world. The fact that few citizens are following the event as closely as their predecessors followed the IGY underscores the need to rejuvenate the nation's Antarctic mentality.

References

'11 Campamentos Antárticos Argentinos en Zona Chilena', *La Unión*, 16 September 1957.

Albert, F. (1901), 'Los Pinípedos de Chile', *Actes de la Société Scientifique du Chili Tome XI*, 220.

Albert, M.R. (2004), 'The International Polar Year', *Science* 303:5663, 1437.

Almeida, A. (1956), 'Reseña Histórica de la Sociedad Chilena de Historia y Geografía', *Revista Chilena de Historia y Geografía* 124, 7.

'Antarctic Claims: Diplomatic Exchanges Between Great Britain, Argentina and Chile in 1951' (1952), *Polar Record* 6:43, 413.

'Antarctic Claims: Recent Diplomatic Exchanges Between Great Britain, Argentina and Chile' (1949), *Polar Record* 5:35–36, 228–41.

'Antarctic Region Claimed by Chile', *The New York Times*, 15 December 1946.

'Antarctic Riches Spur Exploration', *The New York Times*, 10 January 1954.

'Antarctic Rights Reserved by the U.S.', *The New York Times*, 28 December 1946.

'Año Geofísico y la Antártica', *La Union*, 11 November 1956.

'Año Geofísico' (1957a), *La Unión*, 20 May.

'Año Geofísico Internacional' (1957b), *La Unión*, 2 July.

'Argentine Fire Routs British in Antarctica', *The New York Times*, 3 February 1952.

Barrera, H. (1940), 'La Expedición Byrd a la Región Antártica', *Revista Chilena de Historia y Geografía* 96, 285–94.

Barrera, H. (1977), 'Ramón Cañas Montalva', *Revista Chilena de Historia y Geografía* 145, 280–83.

Barrera, H. (1983), 'Los Asuntos Antárticos y la Participación de Algunas Instituciones Chilenas', *Boletín Antártico Chileno* 2, 18.

Berguño, J. (1999), 'El Despertar de la Conciencia Antártica, 1874–1914', *Boletín Antártico de Chile* 18:2, 2.

Berguño, J. (2003), 'Shackleton y Chile', *Boletín Antártico Chileno* 22:1, 42.

Berguño, J. and Canales, R. (2005), 'La Antártica en Punta Arenas, Ayer y Hoy', *Boletín Antártico Chileno* 25:2, 20.

Braun, A. (1974), *Pequeña Historia Antártica* (Buenos Aires: Ed. Francisco de Aguirre).

'British Antarctic Team to Plumb Ice's Depth With Echo Devices', *The New York Times*, 17 January 1955.

Bruggen, J. (1953), 'Informe sobre la Conexión Geológica entre los Andes Patagónicos y los Antárticos', *Terra Australis* 9, 48.

Byrd, R. (1945), Undated Presentation, Byrd Polar Research Center, box 206, folder 7310.

Byrd, R. (1948a), 'Nuestra Marina explora la Antártida', Revista de Marina 331.

Byrd R. (1948b), Letter to Chief of Naval Operations, 7 August 1948, Byrd Polar Research Center, box 206, folder 7328.

Byrd, R.E. (1949), Letter to Commander-in-Chief, U.S. Navy, Pacific Fleet, 10 December 1949, Byrd Polar Research Center, box 62, folder 2799.

Cañas, R. (1945), 'Zona Austral Antártica', *Memorial del Ejército de Chile* 202–3, 514.

Cañas, R. (1956), 'La Antártica', *Terra Australis* 14, 10.

Cañas, R. (1956–57), 'La Antártica y las Proposiciones de la India ante la ONU', *Terra Australis* 14, 3, 10.

'Chile Dio a Conocer Trabajos que ha Realizado en Conferencia Antártica', *La Estrella*, 31 July 1956.

'Chile ha ordenado Construir un Barco Especial Antártico', *La Estrella*, 21 June 1956.

'Chile Inaugura Hoy su Base Científica en el Continente Antártico', *La Unión*, 3 March 1957.

'Chile prestará su Observación Científica sobre la Antártica', *La Estrella*, 15 June 1957.

'Chile Rechazará Toda Proposición de Internacionalización de la Antártica', *La Estrella*, 19 February 1958.

'Chile y el Año Polar Internacional' (1932), *Revista de Marina* 77:448, 399.

'Chile y la Radioactividad Antártica', *Zigzag*, 11 December 1959.

Chilean Antarctic Expedition, 1947 and 1947–48 (1953), *Polar Record* 6:45, 662.

Chilean Decree no. 3310 of December 1902.

Chilean Decree nos. 2074 and 2305 of October 1903.

'Chilean Decree no. 548 of 27 March 1942' (1950), *Polar Record* 5:39, 481.

'Chilean Decree no. 1168 of 15 October 1948' (1950), *Polar Record* 5:39, 482.

'Commandante del Rancagua Habló de la Labor Desarrollada en la Antártica', *La Estrella*, 29 February 1956.

'Consejo de gabinete Acordó Ayer Afianzar los Derechos Antárticos', *La Estrella*, 18 April 1956.

Cook, F.A. (1899), 'Two Thousand Miles in the Antarctic Ice', *McClure's Magazine* 14:1, 4.

Cordovez, E. (1946), 'La Antártida Chilena: El Cuadrante Americano', *Memorial del Ejército de Chile* 212, 85.

Department of State (1955a), Memorandum, 8 November 1955, National Archives, RG 59, Miscellaneous, PPS Office Files, 1955.

Department of State (1955b), Undersecretary of State to Robert Murphy, 23 April 1955, National Archives, RG 59, 399.829/4-2355.

Department of State (1956a), Memorandum of Conversation, 27 July 1956, National Archives, RG 59 399.829/7-2756.

Department of State (1956b), Embassy in London to Bureau of British Commonwealth and Northern European Affairs, 28 February 1956, National Archives, RG 59, 702.022/2-2856.

Department of State (1956c), Embassy in London to Bureau of British Commonwealth and Northern European Affairs, 25 February 1956, Public Record Office, 702.022/2-2856 CS/HHH.

Department of State (1958), Memorandum on U.S. and UK Positions on Antarctica, 28 February 1958, National Archives, RG 59, 399.829/2-2858.

Department of State (1977), *Foreign Relations of the United States 1950 vol. 1, National Security Affairs; Foreign Economic Policy* (Washington: U.S. Government Printing Office).

Domeyko, I. (1847), 'Introducción al Estudio de las Ciencias Naturales', *Anales de la Universidad de Chile*, 143.

'Dos Nuevas Bases Instalará Nuestro País en la Antártica', *La Unión*, 20 January 1957.

Drake, P.W. (1978), 'Corporativism and Functionalism in Modern Chilean Politics', *Journal of Latin American Studies* 10:1, 83, 91.

Duran, S. (1982), 'Aspectos Jurídicos Antárticos', *Boletín Antártico Chileno* 2, 36–41.

Early, S.T. (1949), Letter to Chairman of the U.S. National Security Resources Board, 22 November 1949, Byrd Polar Research Center, box 58, folder 2607.

Edwards, A. (1911), 'Un Nuevo Mapa de Chile', *Revista Chilena de Historia y Geografía* 1, 49–70.

'El Edisto Partió Anoche', *La Estrella*, 5 April 1956.

'El Gigantesco Rompehielos de Estados Unidos que Irá a la Antártica', *La Estrella*, 27 May 1955.

'El Piloto Chileno Luis A. Pardo y su Viaje a la Antártica' (1941), *Revista de Marina* 502, 291.

'El Presidente Llegará hasta la Antártica', *La Estrella*, 21 January 1948.

'El Rancagua y el Lientur Parten esta Noche a la Antártica', *La Unión*, 24 October 1956.

'El Salvamento de los Náufragos del "Endurance" en 1916' (1950), *Revista de Marina* 557, 441.

'The Ends of the Earth' (2007), *Nature* 446, 110.

'En 1958 se Sabrá la Verdad sobre la Riqueza Antártica', *La Estrella*, 17 November 1955.

'En Dos Meses Adelantará su Partida la Flotilla que Irá a la Antártica', *La Estrella*, 14 August 1956.

'En la Antártica Chilena se Construirá un Puente Aéreo', *La Estrella* 28 December 1956.

Escobar, H. (1956), 'El Año Geofísico Internacional', *La Estrella*, 15 September.

Escudero, J. (1953), 'Cincuentenario de la Política Antártica Chilena', *Boletín de la Academia Chilena de la Historia* 48, 74.

Escudero, J. (1984), 'El Decreto Antártico de 1940', en *Anales de la Diplomacia, 1973–1983* (Santiago: Ed. Universitaria).

'Estados Unidos Establecerá Puente Aéreo con la Zona Antártica' (1955a), *La Estrella*, 9 April 1955.

'Estados Unidos Llenará de Satélites el Espacio en el Año Geofísico' (1955b), *La Estrella*, 8 September 1955.

'Estados Unidos Izó su Bandera en el Polo Sur' (1955c), *La Estrella*, 29 December 1955.

'Expedición del Almirante Byrd Llegará a Valparaíso en Abril', *La Estrella*, 3 December 1955.

Evans, E. (1957), *Desafío al Antártico* (Buenos Aires: Editorial Sudamericana).

Fairbridge, R. (1955), 'La Geología de la Antártica y en Especial del Territorio Antártico Chileno', *Revista Terra Australis* 13, 33–51.

Foreign Office (1954), Minute, 28 October 1954, Public Record Office, A 15214/24, FO 371/108767.

Franulic, L. (1956), 'La Antártica, Novia Disputada del AGI', *Ercilla*, 24 October 1956.

Gajardo, I. (1905a), 'Por los Mares Australes: Resumen de las Más Importantes Expediciones Polares Antárticas', *Revista de Marina* 229, 31–39.

Gajardo, I. (1905b), 'Por los Mares Australes: Reminiscencias de la Primera Campaña del "Antarctic" a las Tierras del Rey Óscar II', *Revista de Marina* 228, 641–649.

González, E. (1966), 'Cincuentenario del Salvamento de la Expedición Shackleton', *Revista Chilena de Historia y Geografía* 134, 172.

González-Ferrán, O. (1991), *La Ciencia en la Antártica: Medio Siglo de Política Antártica, 1940–1990* (Santiago: Academia Diplomática de Chile).

'Gran Bretaña Investigará Potencial Científico-Mineral de Antártica', *La Estrella*, 7 June 1955.

Gutierrez, C. and Gutierrez, F. (2006), 'Física: Historia de su Trayectoria en Chile, 1800–1960', *Historia* (Universidad Católica de Chile) 2:39, 477–96.

'Homenaje del Círculo Antártico Chileno a la Memoria del Piloto Luis A. Pardo' (1956), *Revista de Marina* 594, 595.

Honorable Cámara de Diputados de Chile, 2nd Session, 26 November 1946; 2nd Session, 20 October 1953; 36th session, 22 December 1953; 10th Session, 18 May 1955; 13th session, 16 June 1955; 26th session, 12 July 1955; 12th session, 18 April 1956.

Honorable Senado de Chile, 16th and 26th Sessions, 21 January 1947; 18th Session, 22 January 1947; 23rd Session, 13 January 1954; 52nd Session, 17 May 1955; 32nd session, 22 March 1961; 33rd session, 4 April 1961.

Huneeus, A. (1948), *Antártida* (Santiago: Universidad de Chile).

Ihl, P. (1953), 'Delimitación Natural entre el Océano Pacífico y el Atlántico, en Resguardo de Nuestra Soberanía sobre la Antártica y Navarino', *Terra Australis* 9, 47.

'Importante Aparato Científico Instalará la Armada en Punt Arenas', *La Estrella*, 5 March 1956.

'Internacionalización de la Antártica', *La Unión*, 16 February 1958.

'International Polar Year to Advance Understanding of the Earth's Systems' (2007), *Journal of Soil and Water Conservation* 62:2, 27.

Joint Chiefs of Staff (1950), JCS 2070/2, Operations in Polar Regions, 31 January 1950, National Archives, College Park, Maryland, RG 218, CCS 092 Antarctic Area.

Joint Chiefs of Staff (1954), Secret Memorandum for Defense Secretary, 9 July 1954, National Archives, RG 218, CCS 092 Antarctic Area.

Joint Chiefs of Staff (1957), Memorandum for the Deputy Director for Strategic Plans, 2 April 1957, National Archives, RG 218, CCS 092 Antarctic Area.

Joyner, C.C. and Theis, E.R. (1997), *Eagle Over Ice: The U.S. in the Antarctic* (Hanover, New Hampshire: University Press of New England).

Kirwan, L.P. (2001), *Historia de las Exploraciones Polares* (Barcelona: Luis de Caralt Editorial).

Kosack, H. (1955), 'La Explotación de los Yacimientos Minerales de la Antártica', *Terra Australis* 13, 86.

'La Exploración de los Mares Australes por Navíos Españoles Durante el Siglo XVIII' (1998), *Boletín Antártico Chileno* 17:1/2, 2–11.

'La Exploración Rusa Será la Más Grande del Siglo', *La Estrella*, 14 October 1955.

'La Guerra Fría Llega al Continente Antártico' (1956), *Terra Australis* 14, 49.

'La Oceanografía en la Armada de Chile' (1962), *Revista de Marina* 78, 283–85.

'The Last Frontier', *The New York Times*, 5 October 1954.

'Listo el Lugar donde Realizarán Operación Bajo Cero', *La Estrella*, 4 January 1956.

'Llegó Hoy la *Iquique* de Vuelta de la Antártica', *La Estrella de Valparaíso*, 15 April 1947.

López, S. (1975), 'Apuntes Históricos para un Estudio Temático Antártico', *Memorial del Ejército de Chile* 386, 97–98.

'Los Que Abrieron Camino' (1997), *Boletín Antártico Chileno* 16:1, 9.

'Los Rusos Han Empezado a Sobrevolar la Antarctic', *La Estrella*, 10 January 1956.

'Manifestación en Honor de Shackleton y de Pardo' (1916), *Revista Chilena de Historia y Geografía* 24, 195–210.

'Mar Territorial Chileno de 200 Millas Incluye también a La Antártica', *La Estrella*, 16 December 1954.

Martinic, M. (1982), 'Centenario de la Expedición del "Romanche"', *Revista Chilena de Historia y Geografía* 150, 187–89.

Martinic, M. (1999), 'Un Novedoso Mapa Impreso del Siglo XVII Referido al Estrecho de Magallanes', *Anales del Instituto de la Patagonia*, 27, 21–26.

Matte, R. (1903), 'Expedición Al Polo Sur', *Revista de Marina* 35:205, 319.

Miller, E.G. (1950), 'Inter-American Relations in Perspective', *Department of State Bulletin* 22:561, 521–32.

Miller, E.G. (1959), 'Non-Intervention and Collective Responsibility in the Americas', *Department of State Bulletin* 22:567, 768–70.

Mills, C.W (1956), *The Power Elite* (London: Oxford University Press).

Montebruno, J. (1909), 'Reseña del IX Congreso Internacional de Geografía', *Anales de la Universidad de Chile* 125, 301–02.

Moore, J.K. (2001), 'Maritime Rivalry, Political Intervention, and the Race to Antarctica: U.S.–Chilean Relations, 1939–1949', *Journal of Latin American Studies* 33:4, 713–38.

Moore, J.K. (2003), 'Thirty-Seven Degrees Frigid: U.S.–Chilean Relations and the Spectre of Polar Arrivistes, 1950–1959', *Diplomacy & Statecraft* 14:4, 69–93.

Moore, J.K. (2008), 'Particular Generalisation: The Antarctic Treaty of 1959 in Relation to the Anti-Nuclear Movement', *Polar Record* 44:229, 115–25.

Mora, M. (1961), 'El Tratado Antártico', *Anales de la Universidad de Chile* 124, 1813.

Muñoz, J. (1948), 'Antecedentes Geológicos sobre el Sector Pacífico del Continente Antártico', *Revista Terra Australis* 1, 81–87.

'No se Justifican las Protestas en Noruega por Viaje Antártico', *La Estrella*, 21 January 1956.

Olavarría, A. (1965), *Chile entre Dos Alessandri Tomo I* (Santiago: Ed. Nacimiento).

Operations Coordinating Board (1956), 'Progress Report on the U.S. Objectives and Courses of Action with Respect to Latin America', 28 March 1956, Dwight D. Eisenhower Presidential Library, Abeline, Kansas, Office of Secretary of National Security Affairs, NSC Series.

Operations Coordinating Board (1958), 'Report on U.S. Policy toward Latin America', 21 May 1958, Eisenhower Library, White House Office, Office Assistant for National Security Affairs, Policy Planning Sub-Series, Box 18 NSC5613/1, Latin America.

Pacheco, B. (1930), 'Derrotero del Archipiélago de la Tierra del Fuego', *Anuario Hidrográfico de la Marina de Chil* 35, 2.

Pinochet de la Barra, O. (1944), *La Antártida Chilena o Territorio Chileno Antártico* (Santiago: Imp. Universitaria).

Pinochet de la Barra, O. (1976), *La Antártica Chilena* (Santiago: Ed. Andrés Bello).

Pinochet de la Barra, O. (1996), 'Ramón Cañas Montalva, Un Tenaz Precursor Antártico', *Boletín Antártico Chileno* 15:2, 2–4.

Pinochet de la Barra, O. (1997), 'Recuerdos de la Primera Base Antártica', *Boletín Antártico Chileno* 16:1, 7.

Pomar, L. (1903), 'La Expedición Argentina al Polo Sur', *Revista de Marina* 35:207, 437.

'Problema de Importancia Mundial será la Posesión de la Antártica', *La Estrella*, 26 May 1956.

'Quedó Completado el Vuelo entre Punta Arenas y la Antártica', *La Estrella*, 29 December 1955.

'Reclamará EEUU Zona Antártica', *La Estrella*, 28 January 1957.

Reyes, L. (1957), 'El Año Geofísico Internacional', *Memorial del Ejército de Chile* 276, 42.

Rodríguez, E. (1978–1979), 'Participación de Dos Oficiales de la Armada de Chile en la 3° 'Expedición Byrd a la Antártica, 1940', *Boletín de Difusión del Instituto Antártico Chileno* 11–12, 75.

Romero, P. (1985), 'Presencia de Chile en la Antártica', *Memorial del Ejército de Chile* 49, 110.

Romero, P. (1986), 'Apreciación de las Actividades Científicas Antárticas', *Primer Seminario Nacional sobre la Antártica* (Santiago: Ed. Universitaria).

Roosevelt, F.D. (1933), Letter to R.E. Byrd, 7 September 1933, Byrd Polar Research Center, Columbus, Ohio, box 63, folder 2898.

'Se Fijaron los Límites de la Antártica Chilena', *El Mercurio de Santiago*, 7 November 1940.

'Se Ordena la Construcción de un Barco Rompehielos', *La Estrella*, 3 March 1956.

Sievers, H. (1968), 'Una Década de Investigaciones Oceanográficas', *Revista de Marina* 1, 4–20.

Siple P. (1955), Letter to D. Pearson, 28 September 1955, Byrd Polar Research Center, box 58, folder 2603.

Soler, E. (2002), *Viajes de Jorge Juan y Santacilia: Ciencia y Política en la España del Siglo XVIII* (Barcelona: Ediciones B).

Sullivan, W. (1954), 'Soviet Joins 1957–58 World Research', *The New York Times*, 11 December.

Sullivan, W. (1955), 'U.S. Expedition Sails on Antarctic Survey', *The New York Times*, 7 January.

'Temporary Commission on the Liquidation of the Second International Year, 1932–1933' (1950), *Polar Record* 5:40, 621.

'Territorial Claims in Antarctica' (1949), *Polar Record* 5:37–38, 361.

'Torneo Mundial de Geofísica habrá en la Antártica en 1957', *La Estrella*, 2 October 1954.

'Trabajos de Glaciología Chilena del Año Geofísico Internacional' (1962), *Revista Chilena de Historia y Geografía* 130, 417–18.

Transcript of the President's News Conference, *The New York Times*, 18 August 1954.

Tromben, C. (1997a), *Base Prat: Cincuenta Años de Presencia Continúa de la Armada de Chile en la Antártica, 1947–1997* (Valparaíso: Armada de Chile).

Tromben, C. (1997b), 'Se Gesta la Primera Expedición 1946/47', *Boletín Antártico Chileno* 16:1, 3.

'United States Operation Windmill, 1947–48', (1949), *Polar Record* 5:37, 345.

US Embassy in Chile (1950), Telegram to Department of State, 13 July 1950, RG 59, 725.1-2953, National Archives, Collage Park, Maryland.

Urzúa, G. (1968), *Los Partidos Políticos Chilenos: Las Fuerzas Políticas* (Santiago: Ed. Jurídica de Chile).

Vercel, R. (1942), *El Asalto de los Polos* (Santiago: Editorial Difusión).

Vicuña, C. (1915), 'El Territorio de Chile: El "*Utis Possidetis* de 1810"', *Revista Chilena de Historia y Geografía* 18, 151.

Vila, O. (1947), *Chilenos en la Antártica* (Santiago: Ed. Nacimiento).

'Visitas Protocolares Hizo el Jefe del Rompehielos Americano Edisto', *La Estrella*, 2 April 1956.

Würth, E. (1958), *Ibáñez: Caudillo Enigmático* (Santiago: Ed. Del Pacífico).

Zegers, L.L. (1876), 'La Física en la Universidad de Chile: Nota Pasada a Don Alberto Blest Gana, Ministro de Chile en Francia', *Topografía Lahure*.

Chapter 8

Biological Prospecting in the Southern Polar Region: Science-Geopolitics Interface

Sanjay Chaturvedi

Introduction

During the course of the late nineteenth century, Antarctica was subjected to a predominantly state-centric and power-political geopolitics, based on the premise that territory and territorial control necessarily implied more power, prestige and security. Once implicated in the colonial-imperial geographies of exploration and discovery, Antarctica was partitioned into seven pie-shaped claims of territorial sovereignty. The dominant spatial representations of the Antarctic during the 1950s were largely dictated and driven by the Cold War discourses. With the East–West rivalry for power and influence now extending to the south polar region as well, the US proposal to internationalize Antarctica, which eventually culminated in the Antarctic Treaty of 1959 (cited hereafter as the Treaty) was motivated by the broader goals of the containment strategy: keeping the Soviet Union out of Antarctica and its affairs (Dodds 1997).

It was science, especially the IGY, held during 1957–58 which provided a new strategic space to Antarctic geopolitics and laid down the groundwork for the Treaty. The IGY proved to be instrumental in giving rise to the dominant representation of Antarctica as a laboratory for fundamental science and accorded the politicization of Antarctica a totally new, unprecedented direction. Science played a pivotal role in the colonization of Antarctica (Dodds 2006) and the production of scientific knowledge eventually became the key currency of geopolitical influence in the southern polar region. The criteria of 'substantial' scientific activity, to be met by those desiring a consultative status with veto power in the ATS, was skilfully deployed by the major Antarctic powers to decide the inside/outside of the Antarctic domain of responsibility and governance.

During the 1950s and 1960s, it became commonplace for political leaders and scientists alike to describe the Antarctic as either a 'laboratory of science' or a 'continent of peace and science'. The preamble to the Treaty affirms that 'it is in the interest of all mankind that Antarctica shall continue forever to be used exclusively for peaceful purposes and shall not become the scene or object of international discord'. The Treaty (incorporating within its circle of jurisdiction area south of 60° South latitude) prohibits all activity of a military nature and provides for wide rights of inspection to all areas of Antarctica by the Antarctic Treaty Consultative Parties (ATCPs).

Article IV of the Treaty, on account of its innovative containment of contested sovereignty positions, remains central to Antarctic geopolitics. It explicitly declares that 'nothing contained in the present Treaty shall be interpreted as a renunciation by any contracting party of previously asserted right or claims to territorial sovereignty'. The Antarctic Treaty in general and Article IV in particular have escaped a critical scrutiny with regards to the manner in which they 'rewarded' colonial occupation and annexation (Dodds 2006). The legal 'freezing' of territorial claims for the duration of the Treaty (no specific termination date is mentioned) is therefore much more than a carefully crafted diplomatic solution to the thorny issue of claimant and non-claimant states; it protects and promotes a particular vision of the continent anchored in the colonial past.

It was not until the late 1970s that the southern polar region was subjected to significant modern commercial interest. The Convention on the Conservation of Antarctic Marine Living Resources (CCAMLR) was formalized at Canberra in May 1980 and entered into force two years later. The geopolitical consensus underlying Article IV of the Treaty was explicitly reiterated along with its concomitant ramifications for the freedom of the high seas.

By the time the 1988 Convention on the Regulation of Antarctic Mineral Resource Activities (CRAMRA) was signed by the ATCPs, the Antarctic geopolitical equation had significantly changed largely due to the entry of India, Brazil and China into the ATS. The prospects of CRAMRA disappeared eventually when, in May 1989, the Government of Australia announced that it now felt strongly committed to the view that no mining at all should take place in and around Antarctica and a comprehensive environmental protection convention within the framework of the ATS should be pursued instead. French support for the Australian position promptly followed. The consensus within ATS was obviously threatened. In order to enter into force, CRAMRA needed to be ratified by all the countries having territorial claims in Antarctica (Chaturvedi 1996).

The 'u-turns' by Australia and France on CRAMRA brought into question the collective understanding of the ATCPs to abide by the consensus principle, and seriously undermined the capability of the ATS to resolve intra-system conflicts. The announcement on 4 July 1991 of the US decision to sign the Protocol on Environmental Protection to the Antarctic Treaty marked, in a way, the end of the most critical trial of the inner strength and viability of the ATS (Elliott 1994). The Protocol on Environmental Protection to the Antarctic Treaty (hereafter cited as the Protocol) was concluded by consensus on 4 October 1991 at Madrid.

The 1991 Protocol designates Antarctica as a 'natural reserve devoted to peace and science' and binds its present and future signatories to total protection of the Antarctic environment – its intrinsic and extrinsic worth, including its wilderness, aesthetic value and its value as an area for scientific research, especially that which is essential to understanding global environment. It categorically prohibits any activity relating to mineral resources, 'other than scientific research'. The Protocol sets out some basic environmental principles to govern all human activity in Antarctica – be it scientific, tourism related, governmental, non-governmental or related to logistic support. According to the Protocol, activities in the Antarctic Treaty area shall be planned and conducted on the basis of information sufficient to allow prior assessments of, and informed judgments about, their possible impacts on the Antarctic environment and dependent and associated ecosystems and on the value of Antarctica for the conduct of scientific research.

From the early 1980s onwards (marked by both continuity and change in science–politics interplay) Antarctica has been increasingly integrated into global systems and highly capitalized actors and forces of the globalized economy have arrived on the scene. Acting as a major catalyst for this transformation are of course the technological, political and attitudinal transformations in the wider international system (Hemmings 2007).

Within such a context, the key intention of this chapter, on the one hand, is to assess the impact of growing commercialization of Antarctic biodiversity on the legitimacy, authority and effectiveness of the ATS in governing Antarctica for 'peaceful and scientific purposes, and 'in the best interests of humankind'. It examines, on the other hand, the prospects of realizing a post-colonial geography of Antarctica, in the realm of biological prospecting, by bringing about a radical transformation in the ways in which the science–geopolitics interface has been conceived, constructed and deployed over the past fifty years by the Antarctic Treaty Consultative Parties (ATCPs) within the Antarctic Treaty System (ATS) through the mechanism of Antarctic Treaty Consultative Meetings (ATCMs).

Bioprospecting in the southern polar region: Geopolitics, science and market

At the beginning of the twenty-first century, a new revolution in the filed of biotechnology is quite visibly on the horizon. As such, the industries of the future are increasingly targeting the materials and processes in plants, animals and microorganisms. The process of drawing on the chemicals and genetic material of the world's biological resources to provide new feedstock and new modes of manufacture is already under way (Connolly-Stone 2005). This growing industry of biological prospecting has placed most of its attention so far on tropical rain forests and coral reefs. Some of the key sectors pursued in this industry include agriculture, biotechnology, cosmetics, pharmaceuticals and waste management. In 2003 alone, the global biotechnology industry consisted of 4,284 companies (3,662 private and 622 public) in 25 nations, generating $35 billion in annual revenues, and employing some 188,000 people (UNU/IAS 2003). The world's biota represents a source of raw materials that has the potential to replace petrochemicals as an industrial feedstock and to provide novel chemicals for use in drugs and other products (Green and Nicol 2003).

In common parlance, bioprospecting is the evaluation of biological materials for the purposes of assessing their potential utility to the biotechnology industry.

The ultimate goal is to develop marketable biotechnological inventions generally under the protection of patents (Connolly-Stone 2005). In other words, bioprospecting represents a market-driven search for bioactive components in such living organisms as animals, plants, microorganisms (bacteria, microbes) or fungi to develop new commercial products.

Whereas critically conceived, bioprospecting, says Vandana Shiva (2007, 308), 'is a term that was created in response to the problematic relationship between global commercial interests and the biological resources and indigenous knowledge of communities – and to the epidemic of biopiracy, the patenting of indigenous knowledge related to biodiversity'. According to Green and Nicol (2003), bioprospecting comprises the following phases: sample collection; isolation, characterization, and culture; screening for pharmaceutical activity; and development of product, patenting, trials, sales and marketing.

Some analysts have argued that bioprospecting is both progressive and innovative, and carries enormous potential and promise for the development of new products that might prove beneficial to humankind. Consequently, it should be encouraged and commercial enterprises should be rewarded for their investment in such activity in the form of patent rights over the end products. On the other hand, critics argue that claims to patent rights should not be entertained at the cost of freedom of scientific research, and some ways and means must be worked out to share the benefits derived from the commercial use of biological resources.

In global commons areas such as the Antarctic and the high seas, geopolitical considerations of access and ownership, combined with issues of sovereignty and jurisdiction, make bioprospecting and related matters extremely complex. One could possibly identify at least three critical issues that merit special attention and analysis (Green and Nicol 2003). Firstly, commercialization of publicly funded science is likely to impose 'inappropriate' limits on freedom of scientific investigation in both the Antarctic and in the high seas. Secondly, in order to ensure that benefits are shared equitably by the entire humanity in global commons, mutually agreed limitations on ownership rights over biological resources are then required. Finally, consensus will have to be negotiated and sustained by various stakeholders on how best to regulate bioprospecting in areas outside national jurisdiction.

One of the outstanding features of the bioprospecting industry is that research into (and development of) new products often calls for collaborative contractual arrangements between public institutes and the private sector, with the former providing access to collections of samples in exchange for financial support. In addition to public–private consortia, scientists working on a strictly academic project may identify and exploit an organism's valuable use, thus blurring in the process the boundary between scientific research and commercial activity. As pointed in the Report by UNU/IAS (2003, 16) entitled 'the International Regime for Bioprospecting: Existing Policies and Emerging Issues for Antarctica', so far, biological prospecting activities in Antarctica have been carried out by universities, research centres, and biotechnology and pharmaceutical companies, such as the University of Bordeaux (France), the Australian Academy of Technological Sciences and Engineering, Genencor International (multinational), and Merck Sharp & Dohme (multinational). Bioprospecting activities in Antarctica tend to be carried out by consortia comprising a mixture of public and private bodies, making it difficult to draw a clear distinction between scientific research and commercial activities.

There are at least two major reasons behind increasing interest in Antarctic bioprospecting. Firstly, the considerable lack of knowledge surrounding the Antarctic biota provides a unique opportunity to discover and explore potentially valuable new organisms. Secondly, the extremophiles (novel life forms capable of withstanding extreme cold, aridity and salinity) are the most sought after micro-organisms by industry. The application of extremophiles is found in industrial processes such as lipsomes for drug delivery and cosmetics, molecular biology, the food industry and waste treatment (UNU/IAS 2003). In other words, Antarctica offers industry a largely untapped source of valuable extremophile micro-organisms. According to Dr. Hamid Zakri, Director of United Nations Institute of Advanced Studies, Tokyo, 'there is growing evidence that biological prospecting for extremophiles is already occurring and is certain to accelerate in Antarctica and the southern ocean' (cited in Zakri and Johnston 2004).

The growing significance of extremophiles is reflected in the plans for IPY 2007–08 (ATCM 2007). For example, a key question highlighted in the framework for the IPY was: 'How does genetic and functional diversity vary across extreme environments and what are the evolutionary responses underpinning the variation?'. The framework goes on to note that one among the diverse range of activities that will be required for the IPY is marine and terrestrial biological surveying using 'modern genomic methods'. This emphasis is evident in the more than 1,000 research projects that were submitted to the IPY International Programme Office (IPO) to be included as IPY activities. Although this type of research is undertaken for pure scientific reasons, such as the desire to increase the general understanding of Antarctic biodiversity, a significant motivation is the commercial benefits of the outcome of this type of exploration (ATCM 2007, 12).

Among several examples of commercially useful compounds discovered so far, is a glycoprotein, which functions as the 'antifreeze' that circulates in some Antarctic fish, preventing them from freezing in their sub–zero environments. Further research is in progress on the application of this glycoprotein in a range of processes, including increasing the freeze tolerance of commercial plants, improving farm-fish production in cold climates, extending the shelf life of frozen food, improving surgery involving the freezing of tissues, and enhancing the preservation of tissues to be transplanted.

Against the backdrop of several commercial pharmaceutical companies asserting property rights to flora and fauna in Antarctica, as of 2004, more than 40 patents had been granted worldwide on bacteria and other organisms found in Antarctica. In the same year, more than 90 additional patent applications were pending in the United States alone (Stix 2004). Large collections of species are also being created. One example is the Australian Collection of Antarctic Micro-Organisms (ACAM), which houses around 300 species collected from the Antarctic (Green and Nicol 2003). Similar Antarctic bioprospecting activity is being undertaken by public institutes, in partnership with commercial enterprises, in a number of other states (Green and Nicol 2003).

Attracted by such potentially useful discoveries, the private sector has begun to include Antarctic flora and fauna in its product development programmes. Examples of companies' growing interest and activities include a contract signed back in 1995 between the Antarctic Cooperative Research Centre; University of Tasmania, Australia; and AMRAD Natural Products, an Australian pharmaceutical company. According to the contract, AMRAD is given the right to screen some 1,000 Antarctic microbial samples per year in search for natural antibiotics and other human pharmaceutical products. Cerylid Biosciences, an Australian biotechnology company, is also looking for new lead compounds for the development of new anti-cancer and anti-inflammatory medicines. Cerylid bases its discovery work on a biodiversity library containing 600,000 extracts from naturally occurring sources, which include samples of plants, microbes and marine organisms collected in Antarctica. Genencor International, a global biotechnology company (with more than $300 million in revenue in 1999, and over 3,000 owned and licensed patents and applications) also sources materials from Antarctica. One prominent Antarctic scientist is of the view that the private sector has provided at least $1 million funding for Antarctic microbiology and biotechnology since 1997 (UNU/IAS 2005).

Towards an Antarctic bioprospecting policy regime: Imperatives and impediments

The Antarctic Treaty System (ATS) does not directly regulate biological prospecting activities. Nevertheless, there are certain provisions enshrined in the Antarctic Treaty (1961), the Protocol on Environmental Protection (the so-called 1991 Madrid Protocol) and the Convention on the Conservation of Antarctic Marine Living Resources (1982) that might be relevant to bioprospecting. The 1988 Convention on the Regulation of Antarctic Mineral Resources Activities (which eventually was jettisoned in favour of the Madrid Protocol) also contains certain guidelines that might be of some help in developing measures for regulating bioprospecting activities.

At least in the near future, bioprospecting is likely to remain confined mostly to collecting and discovering novel biological resources, an activity that is apparently largely scientific but carries considerable commercial potential and value. Accordingly, bioprospecting activities will fall within the remit of Article

III (a)–(c) of Antarctic Treaty. Under these provisions contracting parties agree that, to the greatest extent feasible and practicable, information regarding plans for scientific programs in Antarctica shall be exchanged to permit maximum economy and efficiency of operations. One could therefore argue that the desire for commercial confidentiality and patents needs to be reconciled with the legal requirements of Article III.

The Antarctic Treaty, under Article IV, states that '[n]o acts or activities taking place while the present Treaty is in force shall constitute a basis for asserting, supporting or denying a claim to territorial sovereignty in Antarctica or create any rights of sovereignty in Antarctica'. This moratorium on sovereignty was a necessary component of the original Treaty in order to neutralize the then-existing unstable political situation caused by the overlapping territorial claims in Antarctica of several treaty nations (Chaturvedi 1996). The fundamental absence of national sovereignty under ATS stands out, in sharp contrast to the 'national sovereignty' approach that forms the basis of Convention on Biodiversity 'bioprospecting' policy as well as that of UNCLOS, except in the high seas and deep seabed (Herber 2006).

Furthermore, in the context of bioprospecting, most research coming from Antarctica requires a considerable investment of resources that are financed mostly by bio-tech industries. As a result, issues such as the ownership of genetic resources and the need to ensure that the resources have been 'legitimately' acquired deserve attention. A lack of clarity about these matters has an important bearing on the plans of companies eying the genetic potential of the Antarctic. In determining ownership and the relevant existing policies governing bioprospecting, jurisdictional issues are therefore of crucial importance. Accordingly, Article VI, states that the Antarctic Treaty applies to the area south of 60° South latitude, including all ice shelves. However, this does not prejudice or affect the rights of any state under international law with regard to the high seas within that area, acquires relevance (UNU/IAS 2003, 9). We will return to this point shortly.

The Madrid Protocol of 1991 and bioprospecting in the Antarctic

The Madrid Protocol of 1991 has a link to bioprospecting via its comprehensive mandate to protect the Antarctic environment and dependent and associated ecosystems. The Protocol designates Antarctica as a natural reserve, devoted to peace and science, and places a moratorium on mineral exploitation (Article 1, Madrid Protocol, 1991). The Protocol sets out a series of environmental principles, which, *inter alia*, stipulate that activities in the treaty area are to be planned and conducted so as to limit adverse environmental impacts, avoid detrimental changes in the distribution, abundance or productivity of species or populations of species of fauna and flora. The Protocol accords 'priority to scientific research, and to preserve the value of Antarctica as an area for the conduct of such research'. This in turn calls for cooperation in the planning and conduct of scientific activities and the sharing of information. The Protocol includes provisions on environmental impact assessment, outlined in Annex I to the Protocol. Thus, prior assessments of the environmental impacts of activities planned pursuant to scientific research programmes, tourism, and all other governmental and non-governmental activates must be carried out.

By virtue of reasoning underlying provisions mentioned above, bioprospecting activities in Antarctica will need to be subjected to an assessment of any potential environmental impacts they may have on the Antarctic environment. It is worth noting in this context that, according to Article 8 of the Madrid Protocol as well as Article VII 5(a) of the Antarctic Treaty, the Environmental Impacts Assessment (EIA) is the responsibility of the state whose nationals undertake the expedition or of the state on whose territory the expedition is organized or proceeds from.

Antarctic Treaty Consultative Meetings (ATCMs) and the challenge of bioprospecting

The question of biological prospecting, or 'bioprospecting' was first discussed at the 25th Antartic Treaty Consultative Meeting with the introduction of a working paper by the United Kingdom (ATCM 2002). At the 27th ATCM held in Stockholm in 2005, the Committee on Environmental Protection (CEP) established 'biological prospecting' as an agenda item (7) and considered the following two information papers presented to the 26th ATCM held in Madrid in 2003: the joint UK/Norway Information Paper on Bioprospecting (ATCM 2003a) and a New Zealand Information Paper (ATCM 2003b).

At the 28th ATCM, held in Stockholm (ATCM 2005), New Zealand and Sweden presented a working paper on biological prospecting in Antarctic. It was pointed out in this document that 'it is unlikely that a bioprospecting activity at the sample collection stage will have any more than a minor or transitory impact, *although this would depend on the particular circumstances*' (emphasis mine). A number of other papers have also raised the question of the relationship of the Convention on Biological Diversity (CBD) with the Antarctic Treaty System, and questions of access and benefit sharing. The 27th Meeting of SCAR held in Shanghai, China, in July 2002, also noted that 'although bioprospecting had been discussed ... previously, this issue requires further attention ... the ATS might need to be extended to include regulation of bioprospecting, and indeed all the provisions of the Convention on Biological Diversity' (UNU/IAS 2005, 20).

At the New Delhi ATCM, held in May 2007, UN Environmental Programme (UNEP) submitted a comprehensive information paper on biological prospecting in Antarctica and pointed out that whereas Antarctic Treaty parties are interested in monitoring the issue, many parties feel they need more domestic engagement, information, analysis and preparation to address this complex issue at the international level (ATCM 2007). Market trends and the momentum in other fora also support the need to be more informed about bioprospecting activities. The

IPY (2007–08) will provide further momentum to scientific studies that contribute to the appeal of bioprospecting and may also result in a new level of interest in the commercial potential of Antarctic biodiversity' (ATCM 2007). According to UNEP,

> Yet no comprehensive or adequate study of Antarctic bioprospecting currently exists, and the reviews conducted thus far have been preliminary and ad hoc in nature. The level of commercial activity that has been brought to the attention of the ATCM to date has been anecdotal. Further research and study is needed to provide a solid informational basis for considering this complex subject, which encompasses scientific and commercial interests, environmental concerns, ethics and equity, and considerations relating to international law and policy, including the adequacy of the Antarctic Treaty System to fully address bioprospecting. (ATCM 2007)

In the light of the above, the analysis to follow examines the relevance and applicability of both the CBD and the 1982 United Nations Convention on the Law of the Sea (UNCLOS) for the purposes of regulating bioprospecting in Antarctica. While examining the possible applicability of the CBD to Antarctica, it is worthwhile noting that, with the exception of the US, the provisions of both treaties bind all ATCPs, who also happen to be the Contracting Parties to the CBD. The difficulty in determining the applicability of the CBD to Antarctica arises from the differing views on whether Antarctica lies outside of the scope of national territories and thus national jurisdiction (UNU/IAS 2003).

The Convention on Biological Diversity and its relevance for the Antarctic

Signed by the representatives of 150 countries at the 1992 Rio Earth Summit, the Convention on Biological Diversity (CBD) is dedicated to promoting sustainable development. It is also conceived as a practical tool for translating the principles of Agenda 21 into reality (Convention on Biological Diversity 2005). The Convention lists three main goals: the conservation of biological diversity, the sustainable use of its components, and the fair and equitable sharing of the benefits from the use of genetic resources. The legal regime it creates is based on the access granted by the states to the components of biological diversity within the limits of their national sovereignty or jurisdiction. According to Article 4 of the CBD on jurisdictional scope,

> subject to the rights of other Sates, and except as otherwise expressly provided in this Convention, the provisions of this Convention apply, in relation to each Contracting Party: (a) in the case of components of biological diversity, in areas within the limits of its national jurisdiction; and (b) In the case of processes and activities, regardless of where their effects occur, carried out under its

jurisdictional control, within the area of its national jurisdiction or beyond the limits of national jurisdiction.

And according to Article 5 of the CBD,

Each Contracting Party shall, as far as possible and as appropriate, cooperate with other Contracting Parties, directly or, where appropriate, through competent international organizations, in respect of areas beyond national jurisdiction and on other matters of mutual interest, for the conservation and sustainable use of biological diversity. (Convention on Biological Diversity 2005)

On the face of it, the above-mentioned provisions do not seem to cover bioprospecting in Antarctica due to the sharp disagreement on the issue of sovereignty in Antarctica. Yet it needs to be pointed out that Article 5 has been used to develop regional efforts to apply the provisions of the CBD. It has also been used as the basis for considering whether the CBD could be applied to regulate the exploitation of marine genetic resources from the high seas and deep seabed.

However, in the case of Antarctica, bioprospecting raises two sets of complex legal issues. The first set of questions relates to the modalities of the activity. Is access to Antarctic biological diversity limited? Is it subject to environment protection requirements? Whereas the second set of questions concerns the results obtained from this activity. How can one reconcile the possible utilization of results with the Antarctic Treaty's requirement that scientific results be made freely available?

As pointed out earlier, the ATCPs have no doubt set up a regime in the form of the 1991 Madrid Protocol that aims at comprehensive protection of the environment. In case the Madrid Protocol is expected to regulate the issues of both access to and use of biological diversity in Antarctica, it would be necessary to define in the first place the competent authority to oversee the bioprospecting activities, and also to specify in unambiguous terms the environmental requirements for this activity. Article 15 (1) of CBD does recognize the sovereign rights of states over their natural resources as well as their authority to determine access to genetic resources, subject to national legislation. This clearly suggests that the final authority to determine access to genetic resources rests with the state. In areas such as the high seas and Antarctic, where there is neither territorial sovereignty nor sovereign rights over resources, access is likely to remain free for all.

Under the Madrid Protocol, all human activity carried out in Antarctica is to be subjected to environmental impact assessments. Interestingly, as of now, this does not include bioprospecting. The EIA is to be based on expected impact of the activity in question on the environment: the greater the impact, the more detailed the evaluation (Article 8). The assumption that bioprospecting can be carried out under environmentally acceptable conditions in the southern polar region needs scrutiny to say the least. Whereas collection of just a few samples is unlikely to adversely affect the environment, this will not be the case if full-scale collection is attempted (Green and Nicol 2003).

UNCLOS III and Antarctic bioprospecting

It has been observed by some scholars that the Southern Ocean legal regime is based upon both the Antarctic Treaty System (ATS) and UNCLOS, and supplemented by international environmental law such as the CBD (Rothwell 2005). The 1982 UNCLOS, which entered into force on 16 November 1994, was adopted in order:

> to establish a legal order for the seas and oceans which will facilitate international communication, and will promote the peaceful uses of the seas and oceans, the equitable and efficient utilization of their resources, the conservation of their living resources, and the study, protection and preservation of the marine environment.

In order to organize and control activities in the 'Area' concerned with seabed minerals, UNCLOS established the International Seabed Authority (ISA), notably with a view to administering its resources (Articles 156–57, UNCLOS). The 'deep seabed' segment of the high seas or 'the Area', as defined in Part XI (Articles 133–40) is the seabed and ocean floor and subsoil thereof, beyond the limits of national jurisdiction. Declared as the common heritage of mankind (Part XI, Article 136), exploration and exploitation in 'the Area' is to be carried out for the benefit of mankind as a whole, irrespective of the geographical location of states. States can neither claim nor exercise sovereignty over 'the Area' and its resources, nor appropriate any part of it (Part XI, Article 137). Here it is important to note that the specific application of this concept under UNCLOS III is to the exploitation of minerals and not to the bioprospecting of biological and genetic resources (Herber 2006).

Furthermore, UNCLOS has established an instrument that authorizes the ISA to regulate mining in 'the Area' (Part XI, Articles 156–58). But as far as bioprospecting is concerned, this body does not have definitive jurisdiction, though it is presently attempting to establish such authority (Part XI, Articles 156–58). Although bioprospecting in the deep seabed is not specifically regulated by ISA at present, it is possible to argue that there is an inextricable factual link between the protection of the deep seabed environment, including its biodiversity, marine scientific research and bioprospecting. For example, the sampling of biological resources may occur in the course of exploration of mineral deposits in 'the Area' (United Nations Environment Programme 2005).

Moreover, several features of the seabed regime outlined in UNCLOS may be extended to (or may become the basis of) a specific bioprospecting policy regime in 'the Area' (United Nations Environment Programme 2005). In any case, the deep seabed area is highly relevant to the formation of bioprospecting policy in

Antarctica since it is an area without national sovereignty. Moreover, a deep seabed (as such) also exists in the Antarctic Treaty area, below 60 degrees South.

Antarctica as a global knowledge commons?

Integral to biological prospecting is the search for knowledge in the domain of diverse biological and genetic resources. Whether or not such knowledge can be described as public good depends largely on the extent to which it is available (or made available) to people at large in a manner that is non-rival and non-exclusive in terms of access as well as consumption (Herber 2006). The defining economic characteristics of a private good, however, are 'rival' in nature and consumed by only one social unit or one individual being at a time. There is indeed considerable evidence to show that in an international political system, characterized by asymmetries in terms of geopolitical clout and technological competence, the ideal of widespread (not to talk of just and equitable) dissemination of knowledge is difficult to realize.

As far as the southern polar region is concerned, it is possible to argue that the Preamble to the Antarctic Treaty of 1961 (and the overall spirit that has dictated and driven the ATS over the past five decades and more) demands that the non-rival (collective) consumption of a public good should prevail. The Preamble to the Antarctic Treaty does recognize that 'it is in the interest of all mankind that Antarctica shall continue for ever to be used exclusively for peaceful purposes and shall not become the scene or object of international discord'.

Accordingly, reference in the Preamble to the effect that Antarctica shall not become an *object* of international discord can also be interpreted to mean that whatever peaceful activities are being undertaken in the Antarctic Treaty area, those engaged in such activities are under an obligation to ensure a 'zero marginal cost' *also for those not directly involved* in the Antarctic Treaty System to enjoy the benefits of such knowledge. In other words, a private good principle is not only in disharmony with the dominant ethos of the ATS, it is also against the principle of global knowledge commons. The production, dissemination and sharing of knowledge (biodiversity in this case) can not be allowed to be guided solely on commercial basis by the private sector in the southern polar region. The vast humanity bordering the Indian Ocean (described by some as the 'Third World Ocean') therefore also has a stake in the biodiversity of the Southern [Indian] Ocean.

It is possible to argue that many of the benefits of knowledge (biodiversity) are 'global' in scope in the sense of non-rival consumption by all world citizens regardless of national locations and identifications. It will be useful for the nascent debate within the ATS on biological prospecting to refine further the classification of knowledge into 'national' and 'global' public good categories. Such a distinction, as Herber (2006) points out, could be made depending upon whether the control of property rights over knowledge falls under the control of state or nation (national public good) or an international legal authority (global public good).

It has also been pointed out by one of the influential Antarctic Treaty consultative members (ATCM 2005b, 3) that, 'the patenting of bioactive substances, resulting from bioprospecting in Antarctica, is not inconsistent with Article III(1) of the Treaty' [exchange of scientific information, observations and results from the Antarctic]. One of the fundamental requirements is that scientific observations and results be exchanged and made freely available. This requirement is qualified by words 'to the greatest extent feasible and practicable'. The proverbial billion dollar question here is this: How do we ensure that the corporations engaged in bio-prospecting willingly adhere to a set of principles and practices that question a culture of secrecy and demand transparency and accountability?

As far as the public-good principle of non-exclusion (a principle reflected in the Preamble to the Antarctic Treaty) is concerned, even if a person does not voluntarily pay for it, he or she should not be excluded from the consumption of that knowledge. Yet, as pointed out earlier, exclusion is possible by manipulation. Both trade secrets and patents can be used to appropriate private property rights. For example, patents associated with medicines obtained from biologically diverse resources may be used (for a specified period of time) to provide exclusive intellectual property rights to an inventor. An important international instrument for protecting such rights, as already discussed in this paper, is the World Intellectual Property Organization (WIPO), which administers 23 international treaties for the protection of intellectual property rights (UNU/IAS 2003).

Thus, what the patents do in practice is that they take out some knowledge out of the public domain and place it in the private domain by assigning property rights to an inventor. And in the process, the possibility of private sector profits is generated 'when such intellectual property rights are created, the public good becomes "mixed" with private good characteristics and, hence, the resulting economic good may be more appropriately termed an "impure" or "quasi" public good rather than a pure public good' (Herber 2006, 140).

Every so-called innovation or invention makes use of previously accumulated knowledge, that is, it draws upon the 'global commons of pre-existing knowledge'. A question that acquires additional significance as well as complexity in the case of the Antarctic is the following: How much of the returns to the innovation should be credited to the innovator and how much would be allocated to the use of the global knowledge commons? In the case of Antarctica, scientific research has historically been characterized by publicly funded and internationally open knowledge, a classic example of global public good.

It is important to note that the principle of global knowledge commons is often evoked (compelling the global bioprospecting industry to pay attention) when private firms prospect for valuable drugs in natural settings (such as the circumpolar North) that involve the cultural geographies and histories of local populations. Such conditions are lacking in the uninhabited southern polar region. Consequently, the global knowledge commons principle throws up unique challenges in the Antarctic.

We might recall briefly that the interest of the 'outsiders' in the icy continent arose as early as the 1980s and somewhat in direct proportion to the origins and evolution of the minerals issue within the ATS. Malaysia was to emerge as the most vociferous and dissatisfied among the critics of the ATS. The Malaysian position reflected, in part or whole, that of most of the developing nations including Antigua, Barbuda, Pakistan, Bangladesh, Cameroon, Cape Verde, Egypt, Ghana, Nigeria, the Philippines, Sri Lanka and Zambia, to name a few. All were critical of the allegedly exclusive nature of the system, the membership of South Africa (no longer an issue in the UN debate), and the distribution of Antarctic resource benefits. From 1984 to 1987, Malaysia and the ATCP's positions over the 'Question of Antarctica' were polarized. Malaysia was bent upon ensuring that Antarctica and its resources were managed in the interests of all mankind and persistent in pursuing the interests of developing countries and demanded that the concept of 'common heritage of mankind' should be applied to Antarctica.

The ATCPs, on the other hand, were quick to oppose the allegation that ATS is anachronistic, discriminatory, harbors colonial territorial claims, is exclusive and thus should be replaced by the common heritage of mankind principle. Whereas the growing membership of the ATS (including the accession to the Antarctic Treaty by India, China and Brazil among others) was underlined to refute the charge of exclusiveness, the 'widely observed principle in international relations whereby those countries primarily engaged in a particular activity are responsible for management and decision making' is being emphasized as 'sensible and working' for the Antarctic. From 2002 onwards, the debate in the UN on the 'Question of Antarctica' can be interpreted as 'constructive engagement'. Malaysia too has increased its direct and indirect Antarctic scientific effort in the late 1990s and early 2000s. Malaysia seems to have accepted the de facto presence of the ATS and has also joined SCAR.

The growing salience of bioprospecting in the Antarctic is likely to raise a number of ethical and equity issues that will not be easy to address and translate into legally binding regulations. An overarching issue of utmost importance here relates to the extent to which biological and genetic resources in the Antarctic are free to be 'owned' by anyone or they should be regarded as the 'common heritage of mankind'. As Alistair Graham (2005, 50) puts it so succinctly in the wider general context,

> We are faced by a genuine 'frontier mentality'. Until such time as the rest of the world catches up, the knowledgeable elites are making up their own rules as they go along, an emerging set of moral principles that has little to do with what we might ordinarily and intuitively expect of ethical behavior to deliver equitable and environmentally sensitive outcomes. Indeed, what evidence we do have indicates that these elites will fight very hard indeed to forestall the imposition

of a more customary 'ethical' framework by the wider community as awareness dawns upon them as to what is at stake.

Conclusion

A closer look at market trends suggests continued growth of the biotechnology industry, including in the pharmaceutical, enzyme, cosmetics, chemistry and agricultural sectors. And with increasing research on Antarctic micro-organisms, the knowledge base is steadily improving, ably assisted by technological improvements. As pointed out in this chapter, a major goal of the IPY is to further develop this base. These growing markets are likely to fuel the demand for novel genetic resources and are likely to result in a much enhanced pace and scope of interest in Antarctic microbes.

Looking ahead, the challenge is to negotiate an Antarctic bioprospecting regime built around the long-established Antarctic scientific tenets of public funding and international openness. This would demand not only sharing of information and knowledge among those (state and non-state actors) engaged in biological prospecting in utmost transparency (something easier said than done) but also the adoption of stringent environmental regulations and impact assessments. Such a pursuit might even compel the ATCPs to rethink and modify some of the provisions of the Madrid Protocol and its annexes.

Should private companies be allowed to profit from species unique to the Antarctic as yet another 'peaceful' use of Antarctica and the Southern Ocean? This is a difficult but unavoidable question. It needs to be noted that ever since the Antarctic Treaty came into force in 1961, the term 'peaceful uses' of Antarctica has been steadily expanded beyond scientific research to include commercially driven activities such as fishing in the Southern Ocean and tourism.

On the one hand it is possible to argue that, provided there is a proper regulatory regime in place, biological prospecting could be treated like other activities such as fishing and/or tourism, provided it does not harm the environment and benefits the humankind as a whole. On the other hand, one could argue with equal conviction that since Antarctica is set aside under the 1991 Environment Protocol to the Antarctic Treaty as a protected area dedicated to open science and environmental protection, to allow a free-for-all on bioprospecting is a violation of these values, including the long-standing imperative within the ATS of sharing all scientific information freely. Let us turn to a more detailed discussion of the complex issues involved in this debate.

The absence of clear rules governing the use of genetic resources from Antarctica restricts use of these resources and affects the behavior of stakeholders in significant ways. For industry, for example, the uncertainty about the use and ownership of samples inhibits their support for Antarctic research. For scientists, a lack of clear protocols with regard to exchanging information arising from commercial activities inhibits their ability to work with companies and adapt at the same time to the changing nature of basic research around the world. For governments, it is likely to prove difficult to negotiate how benefits of commercially orientated research could be adequately shared in the best interests of humankind as a whole.

As pointed out by Vandana Shiva (2007, 309), 'nature's biodiversity and diversity of knowledge systems are undergoing a major process of destabilization with the expansion of patents and intellectual property rights into the domain of biodiversity via the Trade-Related Aspects of Intellectual Property Rights (TRIPs) agreement of the World Trade Organization (WTO). The whole notion of TRIPs has been shaped by the objectives and interests of trade and transnational corporations'.

Although the nature and extent of the physical impact of bioprospecting on the Antarctic eco-systems and biodiversity is being currently addressed by the ATCPs, the task of putting into place (through consensus-based negotiations) a sound legal–political arrangement (one that resists, restricts and regulates the commercialization of polar biodiversity, in harmony with the principles of equity and fairness) is much more complex than often assumed by both the scholars and policy makers. Indeed, developing sound and sustainable measures on bioprospecting in Antarctica would require some basic conceptual agreement on the overall goals of any regulation and the type of management system that is desirable, practical and – most importantly – equitable. Since bioprospecting is an activity with potentially both environmental and resource implications, the Antarctic Treaty parties need to work out, sooner than later, a more comprehensive policy position, if not a regulatory framework. Coordination with other international legal fora is seemingly an inevitable aspect of the formation of a comprehensive Antarctic Bioprospecting Policy regime.

The role of science as the key currency of geopolitical influence in the southern polar region is likely to be reinforced by bioprospecting in the Antarctica. It will continue to be deployed for the purposes of both reenforcing and undermining territorial claims. In the light of burgeoning research in cultural histories of science, as noted by Klaus John Dodds (2006) a far more serious and systematic research is needed to investigate how various scientific discourses, and the practices flowing from them, contribute to the consolidation and exercise of geopolitical power in the Antarctic. To quote Dodds (2006, 62–63).

> As has been widely noted, the role of science is critical in facilitating the Antarctic Treaty System (ATS) … However, social scientists have devoted less attention to how these scientific practices contribute to the consolidation of geopolitical power. This might require, for example, a more detailed analysis of the actual practices surrounding the annual Antarctic Treaty Consultative Meetings and the Scientific Committee on Antarctic Research (SCAR). As the Treaty parties have improved accessibility for observers, including ASOC, so the opportunities to disseminate scientific knowledge have improved even if the knowledge production is concentrated within particular states and regions. This is important because

scientific endeavour has enabled states such as the United States and English speaking allies – Britain, Australia and New Zealand – to shape political agendas …Within the cohort of Antarctic Treaty Consultative Parties (ATCPs), there are important differences relating to capacity and subsequent political influence.

Antarctica is now increasingly exposed to global forces and the ATS appears under pressure in terms of exerting its legitimacy, authority and effectiveness. Commercial competition is beginning to displace scientific cooperation as the driver of policy into the region. This is not to suggest that the ATS is going to collapse in the wake of mounting pressures by corporate globalization in near future, but it does appear to be in relative decline and power is slowly but surely shifting from state to non-state entities, and particularly to commercial enterprises and interests.

References

ATCM (2002), 'Biological Prospecting in Antarctica', XXV ATCM/WP-043 (Submitted by United Kingdom).
—— (2003a), 'The International Regime for Bioprospecting: Existing Policies and Emerging Issues for Antarctica', XXVI ATCM/IP 057 (Submitted by Norway and the United Kingdom. Written by S. Johnston and D. Lohan, UNU/IAS).
—— (2003b), 'Bioprospecting in Antarctica: An Academic Workshop', XXVI ATCM/IP 047 (Submitted by New Zealand).
—— (2005), 'Biological Prospecting in Antarctica', XXVIII ATCM/WP13 (Submitted by New Zealand and Sweden).
—— (2007), 'Biological Prospecting in Antarctica: Review, Update and Proposed Tool to Support a Way Forward', XXX ATCM/IP 67 (Submitted by UNEP).
Chaturvedi, S. (1996), *The Polar Regions: A Political Geography* (Chichester: John Wiley & Sons).
Connolly-Stone, K. (2005), 'Patents, Property Rights and Benefit Sharing', in Hemmings and Rogan-Finnemore (eds).
Convention on Biological Diversity (2005), 'Proposals by Switzerland Regarding the Declaration of the Source of Genetic Resources and Traditional Knowledge in Patent Applications', Item 4 of the Provisional Agenda. *Third Meeting of Ad Hoc Open-ended Working Group on Access and Benefit-sharing*, Bangkok, 14–18 <http://www.biodiv.org/doc/meetings/abs/abswg-03/information/abswg-03-inf-07-en.pdf>, accessed 1 October 2008.
Dodds, K.J. (1997), *Geopolitics of Antarctica: Views from the Southern Oceanic Rim* (Chichester: John Wiley & Sons).
—— (2006), 'Post-Colonial Antarctica: An Emerging Engagement', *Polar Record* 42:220, 59–70.
Elliott, L.M. (1994), *International Environmental Politics: Protecting the Antarctic* (London: Macmillan).

Graham, A. (2005), 'Environmental, Ethical and Equity Issues', in Hemmings and Rogan-Finnemore (eds).

Green, J.A. and Nicol, D. (2003), 'Bioprospecting in Areas Outside National Jurisdiction: Antarctica and the Southern Ocean', *Melbourne Journal of International Law* 4:1, 76–111.

Hemmings, A.D. (2007), 'Globalization's Cold Genius and the Ending of Antarctic Isolation', in Kriwoken, Jabour and Hemmings (eds).

—— and Rogan-Finnemore, M. (eds) (2005), *Antarctic Bioprospecting*. Gateway Antarctica Special Publication Series 0501 (Christchurch: University of Canterbury).

Herber, B.P. (2006), 'Bioprospecting in Antarctica: the Search for a Policy Regime', *Polar Record* 42:221, 139–46.

Kriwoken, L.K., Jabour, J. and Hemmings, A.D. (eds) (2007), *Looking South: Australia's Antarctic Agenda* (Sydney: The Federation Press).

Madrid Protocol on Antarctic Environment Protection (1991), <http://www.ats.aq/documents/recatt/Att006_e.pdf>, accessed 1 October 2008.

Rothwell, D. (2005), 'Southern Ocean Bioprospecting and International law', in Hemmings and Rogan-Finnemore (eds).

Shiva, V. (2007), 'Bioprospecting as Sophisticated Biopiracy', *Signs: Journal of Women in Culture and Society* 32:21, 308–13.

Stix, G. (2004), 'Staking claims – Patents on Ice: Antarctica as a Last Frontier for Bioprospectors – and their Intellectual Property', *Scientific American*, 1 May <http://www.sciam.com/article.cfm?articleID=0007671B-A73E1084A73E83414B7F0000>, accessed 1 October 2008.

United Nations Environment Programme's Governing Council (2005), *Report of the Executive Director on State of the Environment and Contribution of the United Nations Environment Programme to Addressing Substantive Environmental Challenges* December 22, UNEP/GCSS.IX/10 <www.unep.org/GC/GCSS-IX/Documents/K0584611GCSS-IX-1add1.pdf>, accessed 1 October 2008.

United Nations University Institute of Advanced Study (UNU/IAS) (2003), *Report on the International Regime for Bioprospecting: Existing Policies and Emerging Issues for Antarctica*, Tokyo <http://www.ias.unu.edu/binaries/UNUIAS_AntarcticaReport.pdf>, accessed 1 October 2008.

—— (2005), *Report on the Bioprospecting in Antarctica*, Tokyo <http://ecosystemmarketplace.net/documents/cms_documents/antarctic_bioprospecting.pdf>, accessed 1 October 2008.

World Trade Organization (1994), TRIP Agreement, <http://www.wto.org/english/tratop_e/trips_e/t_agm3c_e.htm#5>, accessed 1 October 2008.

Zakri, H. and Johnston, S. (2004), *Report: Accelerate Global Agreement to Oversee Exploitation of South Pole 'Extremophiles': Ownership of Genetic Materials Environmental Consequences in Question as 21st Century Bio-prospecting Gets Underway in Antarctica* <http://www.unu.edu/news/extremophiles.html>, accessed 1 October 2008.

Chapter 9

Three Spirals of Power/Knowledge: Scientific Laboratories, Environmental Panopticons and Emerging Biopolitics

Monica Tennberg

Some of the scientific changes which have taken place in the polar regions are reflected in the logos of the IGY (1957–58) and the IPY (2007–08). The IGY logo represents a time of scientific expeditions, technical advancement and growing political interest in the Antarctic in the late 1950s. In the logo, the globe is tilted towards the Antarctic and a satellite orbit circles the globe. In comparison, the IPY logo represents a totally different view of the polar regions in the world of science; it depicts the growth of human impacts on the world environment as well as the increasing role of the human dimension of research in the scientific program of the IPY. In the fifty years that have passed between IGY and IPY, science and knowledge have played important but different roles in the political development of the regions. As Jabour and Haward conclude in Chapter 5, 'we regard the Antarctic as an exemplar of the science–policy interface … It is the same in the Arctic, where decisions rely on both scientific input and state will'.

Both the IGY and the IPY are interesting 'events' to study, not necessarily as a part of continuous historical developments but as possible points of changing directions and new beginnings. As a method, Foucault (1991a, 76–77) 'eventalization' rediscovers the connections, encounters, supports, blockages, forces and strategies which at a given moment establish what subsequently counts as being self-evident, universal and necessary. In simple terms, the aim is to show the multiplicity of processes behind an event, such as spirals of knowledge and power, confront and support each other (see also Michon 2005). Knowledge and power, as Michel Foucault (1980, 98) suggests, are closely linked; they both support each other and conflict with each other. Power is not only restrictive, but also productive and effective. Power relations are everywhere, as Foucault's thinking suggests, even in such remote areas as the polar regions. Foucault's (2007, 17) analysis of power focuses on 'where and how, between whom, between what points, according to what processes, and with what effects, power is applied'.

Power relations can be understood in terms of a continuum. At one end, there are the relations of power as sovereignty and at the other end there are the relations of power as discipline. Power as governmentality is somewhere in between the two ends. Governmental power is not constrained to state power only, but includes different forms of power and use of power within states and between them (Sending and Neumann 2006; Merlingen 2006). Using Foucault, Singer and Weir (2008) have claimed that different forms of power, sovereignty, discipline and government, co-exist and interact. These powers are heterogeneous, with multiple principles of intelligibility and they are interactively related. These three different forms of power work in the polar regions to produce a dynamic complex of knowledge, science and politics.

Knowledge, then, is best understood, as Foucault thinks, as discourses made of subjects, objects, strategies and enunciative modalities. These discourses are connected to non-discursive practices (Foucault 1972, 107). Different forms of power require different kinds of knowledge: sovereign power needs knowledge to maintain its power in terms of resources; disciplinary power is based on detailed knowledge of the disciplined and their practices; and, finally, governmental power requires knowledge about the governed, the population and about the milieu in which they live. The systems of knowledge production in the polar regions aim to fulfill these different needs for knowledge. Knowledge is a source and an effect of power relations, but it is also often contested. Knowledge is contested in at least two ways: what constitutes knowledge about the polar regions and who knows polar regions the best. Knowledge effects are produced by the struggles, confrontations and tactics of power (Foucault 2007, 18).

From this starting point, the legacies of scientific and political cooperation are effects of these power relations and their workings; the polar regions are established as 'scientific laboratories' and part of global 'environmental panopticons', and a system of surveillance for environmental research as highlighted in the tradition of IPYs in polar regions. Nonetheless, as Foucault suggests, power relations are always challenged, and effects of power are contested in various ways. In the polar regions, new ways of understanding and using knowledge questions the existing modes of governing and their knowledge base.

Re-territorializing the polar regions

The territories of the polar regions and their natural resources have attracted national interests for centuries. The reason for the state is its own power and its enhancement through territory (Foucault 2007, 28). In the history of polar expeditions, national and scientific interests were in many cases linked, as Shadian notes in Chapter 2, 'scientific exploration has been one the hallmarks serving as a means for securing and expanding sovereign authority over territory' in the polar regions. Conversely, the recent history in terms of sovereignty is very different in the two regions, as pointed out by Jabour and Haward (Chapter 5): 'Paradoxically, the Antarctic has a well-established, coherent regional management regime with legal personality, whereas the Arctic, with its population, resource exploitation and prominent strategic profile, does not have the same cohesive legal regime'. Through scientific cooperation, the polar regions were re-territorialized after the

Second World War. They were territorialized from objects of national claims of land and international disagreement to objects of scientific cooperation as a 'scientific laboratory'.

Particularly in the Antarctic, this re-territorialization took place early, in the late 1950s. Through the IGY 1957–58 and the 1959 Antarctic Treaty, Antarctica was made into a laboratory for fundamental science, and conflicting claims of territorial sovereignty were frozen. Rothwell (Chapter 6) describes the ATS as a special formula under which the states could join a single regime 'without compromising their position on the status of sovereignty claims, or potential sovereignty claims'. Furthermore, according to Rothwell, the Antarctic Treaty has proven 'invaluable' in the continuing promotion of Antarctica as a 'continent of science' in the tradition of the IGY. Rothwell concludes that 'the effect of the IGY was to demonstrate the importance of Antarctic science and the virtue of international cooperation between Antarctic scientists'. According to Jabour and Haward (Chapter 5), the Antarctic Treaty 'reinforced the role of science and scientific collaboration, first established through polar discovery, exploration and the coordinated effort of the first and subsequent IPY or IGY'.

In the Arctic, the idea of the region as an international scientific laboratory gained acceptance much later. Nilsson describes in her chapter (Chapter 1) how Cold War politics created major obstacles for scientific cooperation. Despite the political constraints, 'the wish for pan-Arctic research cooperation remained alive within the polar research community and among non-governmental scientific networks'. In the late 1980s discussions led to the creation of the International Arctic Science Committee (IASC) in 1990. This was paralleled by 'a surge in political diplomatic activity' leading to the Declaration on the Protection of the Arctic Environment and the creation of the Arctic Environmental Protection Strategy (AEPS) in 1991. Arctic environmental cooperation was institutionalized by the establishment of the Arctic Council in 1996. The concern over the Arctic environment also produced a number of other international initiatives during the 1990s, for example, the Barents Euro-Arctic Council and the Northern Forum.

Arctic cooperation has led to many assessments, reports and guidelines dealing with various environmental concerns in the region. In the Arctic Council, indigenous peoples' organizations not only have a formal role as 'permanent participants'; they also take part in scientific assessments. Indigenous peoples have a role as 'knowledge providers' in the Arctic. Control over how the Arctic is framed by science is also part of indigenous peoples' claim to self-determination and autonomy. Arctic indigenous peoples have been able to use the process of political and scientific environmental framing of the Arctic to create a role for themselves as local experts. This challenge to Western scientific practices is, according to Nilsson (see Chapter 1), 'part of a more general trend', however, it 'is particularly visible in the Arctic, possibly because indigenous peoples are fairly well organized and also have the Arctic Council as a political forum to act in'. It is certainly an important part of the political context for the emphasis on human dimensions in the 2007–08 IPY. Nilsson concludes in Chapter 1 that 'in polar research it has become especially pronounced in the Arctic where scientists and nation states are no longer the sole legitimate transnational actors and scientists no longer the only legitimate knowledge providers, but where political and scientific communities as accompanied by indigenous peoples'.

The issues of territory and sovereignty have recently emerged onto the political agendas of polar regions, though perhaps it never left. In the Arctic, climate change and its promises of easier access to natural resources in the region has inspired the Arctic states to renew and strengthen their territorial claims. The region around the North Pole is currently divided between Canada, the USA, Russia, Norway and Denmark, but there are many unsettled issues and disagreements about borders and boundaries. In 'the new race to the North Pole', Russia and other Arctic countries have become active in taking action to make claims or strengthen existing claims on the region. In August 2007, a Russian mini-submarine dropped a Russian flag on the ocean floor at the North Pole as a symbolic claim of the polar region's oil and minerals ('Russia Plants', 2007). More recently, the Russian Federation decided it is necessary to have clear borders in the Arctic (Lapin Kansa 2008). As Shadian notes (Chapter 2), 'the reaction [to the Russian flag ceremony] which ensued was a reawakening to the strategic importance the Arctic for the eight Arctic countries. Merely in terms of science, immediate responses included joint research expeditions from the Arctic countries including Sweden, Denmark and the United States. ... The Russian event itself, it could be argued, did as much to advance both public and scientific interest in the polar regions as did the kickoff for the fourth IPY'.

The issue of sovereignty has not disappeared from the Antarctic political agenda either. In the Antarctica, existing arrangements for science/policy interface are described by Jabour and Haward (Chapter 5) as problematic in the sense that 'the somewhat romantic notion of giving scientific research pre-eminence over politics has been overtaken by the political realities of the Antarctic regime. These include claimants wishing to protect potential benefits from claimed territory at some point in the future, with claimant parties using scientific research as a shield of legitimacy'. In Antarctica, Chaturvedi considers 'the legal "freezing" of territorial claims for the duration of the Treaty (no specific termination date is being mentioned) ... protects and promotes a particular vision of the continent anchored in the colonial past' (Chapter 8). This issue of sovereignty still haunts the ATS as Woppke concludes (in Chapter 7): 'from the earliest days of independence through participation in the IGY, Chile's Antarctic mentality remained fixated on national sovereignty. Chileans still believe that the nation's permanent occupation provides one of the strongest bases for its rights, not scientific cooperation'. The issue of sovereignty is part of the everyday politics of the ATS, where 'the scientific institutions established by instruments of the ATS are essential from a practical perspective, but they are primarily advisory in nature. While their scientific research might figure prominently in the rhetoric of the regime, the Realpolitik is that the power', as Jabour and Haward state (in Chapter 5), 'belongs to the diplomats and policy makers'. Knowledge and power have reinforced each other in the polar regions, but they have also set new restraints as Rothwell points out (Chapter 6); 'the only real limitation that has been imposed upon the freedom of scientific research throughout the duration of the Treaty has been through additional instruments and mechanisms adopted under the ATS'.

Setting up environmental panopticons

The history of the IPYs can be seen as a process of disciplining the regions through the means of science. Discipline structures a space, a panopticon, an organization of space where each person is seen, but he does not see; he is the object of information, never a subject in communication (Foucault 2007, 39–40). Disciplinary power makes the environment and the people in it objects of surveillance. Deacon (2002) makes a distinction in understanding the role of disciplinary power in history: modern history is one of 'disciplinary power, but not disciplined'. Modern disciplinary power is 'a product of gradual but discontinuous and contingent convergence among multiplicity of localized and comparatively minor, but infinitely productive and permanently contested practices'. The polar regions are part of a global environmental system; a system developed for the needs of weather forecasting and climate science and which is closely linked to the history of IPYs in the polar regions.

In Chapter 1, Nilsson describes how the IGY became a starting point for looking at planet earth as a whole. While there was an early connection between climate science and the Arctic, the motivation for international coordination came from more immediate concerns, namely a need to predict weather, which led to collaboration among national meteorological offices. The early development of climate science and its relation to polar research shows how individual scientists and scientific networks with an interest in the ice ages became increasingly organized and international in their orientation, even though colonial and nationalistic tendencies dominated scientific cooperation in the Arctic region. Nilsson describes how the breakthrough for international climate science came via a wish to use new satellite technology to gather data about weather conditions high in the atmosphere. Nilsson summarizes the post–World War decades (in Chapter 1), as 'formalized governmental and non-governmental cooperation [which] created networks of people and technologies that emphasized the earth as a system'. Emerging environmental panopticon – a scientific gaze – over polar regions was, as Nilsson describes, part of 'the dawn of a cultural era that encompassed ideas of both the destructive power of human technology and viewing the earth as a whole' (see Chapter 1).

Tangible and practical products of this new global thinking are the polar research stations that Wråkberg discusses in Chapter 4. The IPY was seen as 'a unique and compelling opportunity to develop sustained observing systems at both poles'. The IPY is based, according to Wråkberg, on 'the idea that synoptic internationally coordinated field data collection undertaken by professional scientists at well-organized polar stations will produce knowledge of superior quality and greater interest to science than individual data-gathering based on small scale logistics'. The IPY field stations are related to one or more distant centres of calculation and to networks for exchanging scientific information and negotiating credibility. A field station 'is one of many sites of scientific knowledge production and related to other such institutions of social interaction, research and re-production: universities, institutes, academies, observatories, laboratories, testing ranges, excavation areas, field camps, research vessels and expeditions' (Wråkberg in Chapter 4).

The history of IPYs is closely connected to the institutionalization of scientific international cooperation through its establishment of scientific bodies to assist political decision-making, such as the Scientific Committee on Antarctic Research (SCAR), and to the establishment of international scientific organizations, such as the World Meteorological Organization (WMO), and International Council of Scientific Unions (ICSU). However, the politics of knowledge is lively even among the established bodies of knowledge such as SCAR, described by Jabour and Haward in Chapter 5. The discussion focuses on SCAR and its role as a knowledge producer in the Antarctic Treaty System (ATS). While SCAR was not mandated expressly with the task of scientific coordinator and advisor within in the ATS, it inherited it as a legacy of its IGY responsibilities. SCAR considers itself 'the primary source of independent scientific advice' to the ATS although, 'its supremacy in this role has been questioned in recent times'.

The history of the IPY is one of research expeditions and stations. In the present IPY, as Rothwell reports in Chapter 6, there has not been an expansion of existing Antarctic scientific research bases and stations. The ATS, especially the Madrid protocol, has set 'significant limitations' on the building of such facilities in Antarctica. Furthermore, growing awareness and concerns over environmental impacts of such stations has limited enthusiasm for establishing new research facilities in Antarctica. In the Arctic, old stations have been reopened such as the Nordic research station in Kinnvika, in Spitsbergen. According to Wråkberg (Chapter 4), the re-opening of the Kinnvika station turned it into a symbol, through its value as historical novelty to most Scandinavians, its fair state of preservation, and through actively appropriating and shaping positive meanings around the IGY station. It was a nostalgic call to new generations of polar researchers and more importantly, for more research funding.

In further connection to IPY, plans for new stations have been announced, such as those by the Canadian government. The aim of the Canadian plans is to establish a new research station in the Canadian Arctic region, thereby strengthening its territorial claim and sovereignty over the area ('Arctic Research Station' 2008). It is difficult, however, to think that if the new Canadian research station were to be established, that it would follow the traditional logic of building research stations in remote areas with little contact with their surroundings, especially with indigenous communities. In the context of lively North American discussions about indigenous knowledge and knowledge producers, there are pressures to re-think the idea of a research station and its working practices with the local communities in new, constructive ways. As the memo of the plan notes:

> ... the peoples who have in lived in Canada's far north for countless generations have unique histories that are closely linked to sustained, flexible relationships with the land, coasts, oceans, and sea ice which provide vital sources of sustenance, shelter, and identity for northern peoples. These relationships and associated values and traditions constitute a knowledge basis for Arctic citizens to engage publicly with science and technology through participatory approaches to research, consultation decision-making and environmental regulation. ('Canadian Research Station' 2008)

Shadian describes in Chapter 2 how, in the Arctic, indigenous knowledge has become 'a significant means by which indigenous peoples have legitimized a stake, and claims to, control over local research and development practices'. The way the new Canadian research station will be formulated and operationalized will have to take into account these contested claims of knowledge and new modes of knowledge production. Moreover, and in contrast to the 'global gaze' to the Arctic climate and its impacts regionally, indigenous participation also emphasizes local impacts of climate change and how changes in the physical environment interact with social, cultural, and political processes. Their role as knowledge providers, as Nilsson describes in her chapter (Chapter 1), is 'closely intertwined with a critique against globalism, as defined by western science, as a sole legitimate framing of environmental problems'. Shadian assumes (in Chapter 2) that 'as the consequences of this IPY begin to unfold, indigenous knowledge will leave lasting legacies for the way in which the future of all Arctic scientific exploration will proceed and Arctic science (namely climate change) will be understood more broadly'.

Emerging biopolitics in polar regions

The politics of governmentality is called 'biopolitics' by Foucault. All institutions and practices concerned with exploiting, managing and protecting the environment are expressions of biopolitics. Biopolitics focuses on, as Mitchell Dean (1999, 99) adds, the social, cultural, environmental, economic and geographic conditions under which humans live, procreate, become ill, maintain health or become healthy, and die ... it is concern with the biosphere in which humans dwell. Through governmental power, life itself – the vital reality of a people – becomes the responsibility of political authority (Rose 2001).

The essential issue of government will be the connection between political practice and economy. The aim of the state is not to maintain or strengthen its sovereignty nor establish discipline, but to govern the population, to an end convenient to the governed (Foucault 1991b, 94–95). This implies a plurality of aims, where the state has to ensure that the greatest possible quantity of wealth is produced, that the people are provided with sufficient means of subsistence and that the population is enabled to multiply etc.

Emerging governmental practices of power in polar regions means that the polar regions will be part of the working of the global economy increasingly and more intensively than before. Chaturvedi describes the developments (in Chapter 8) that 'from early 1980s onwards (marked by both continuity and change in science–politics interplay) Antarctica has been increasingly integrated into global systems and highly capitalized actors and forces of the globalized economy have arrived on the scene'.

In the Antarctic, the issue of bioprospecting challenges the understanding of knowledge, the role of science and existing modes of governing. Bioprospecting is the evaluation of biological materials for the purposes of assessing their potential utility to the biotechnology industry, where living organisms, such as animals, plants, microorganisms (bacteria, microbes) or fungi are developed into new commercial products. Chaturvedi points out that 'in global commons areas such as the Antarctic and the high seas, ... considerations of access and ownership, combined with issues of sovereignty and jurisdiction, make bioprospecting and related matters extremely complex' (Chapter 8).

The issue of bioprospecting is also part of the IPY scientific agenda as pointed out by Chaturvedi. There are several questions and challenges in connection to bioprospecting and its governance in the Antarctic. Chaturvedi lists them in Chapter 8: Firstly, what effects will the commercialization of publicly funded science will have on freedom of scientific investigation in both the Antarctic and in the high seas. Secondly, how benefits involved in bioprospecting will be shared among the participants in the ATS and by the humanity at large. Finally, how to best regulate bioprospecting with the existing legal frameworks and beyond them. Rothwell (in Chapter 6) adds to these legal complications with the governance of bioprospecting questions such as: 'Is this an activity which is best regulated under national laws, or is it appropriate for international legal regulation? If international responses are adopted, are general framework provisions in order or is there a need for detailed laws and if so over which areas should these laws apply – the high seas only, or does it extend to both the oceans and terrestrial areas?' Rothwell concludes that the issue of bioprospecting 'holds the seeds of possible dispute, especially if distinctions are sought to be drawn between the freedoms associated with the exploitation of genetic resources and the limitations on natural resource exploitation more generally, especially Antarctic minerals'.

Another question is the connection between knowledge and systems of governance in the polar regions. In Chapter 3, Huebert claims that 'science by itself does not lead to better international cooperation. It may be hoped for, but it does not occur on its own. How new scientific information can be used to help improve cooperation is not well understood. Yet there does not seem to a willingness to support the social sciences to understand this process'.

A new development in the current IPY is research projects by private companies. Shadian gives two examples of such Arctic IPY projects in Chapter 2. In this sense the fourth IPY marks a new beginning for scientific cooperation: 'While this IPY is said to be unique in many ways including the introduction of the human dimension and social sciences, it also marks a distinct break from IPY tradition through the formal inclusion of private industry in IPY research proposals'. Shadian explains that that formal private investment in polar science is highly non-traditional, although during the IGY, private industry was actively participating in the research. Whereas the IGY was about public funding for military science, in the current IPY cases private industry is funding public science. Shadian's chapter questions what the implications of these changes are. 'Is bringing in more participants into research the democratization of science or is it bringing science out of the "official" public into the "private" sphere? Is private funding of science undermining public input into science or are the spaces of science shifting and therefore redefining the definitions traditionally contained within the conceptual boundaries of public and private science?'

The emerging biopolitics in polar regions set new challenges for governance of these regions. The current global mode of governmentality is neoliberalism, which stands for a complex set of discourses and institutional practices that have spread worldwide. The changing dynamics of private and public science, companies and policy in polar regions is the workings of governmental power. Defining the role of the state is central to it: 'It is the tactics of government which make possible the continual definition and redefinition of what is within the competence of the state and what is not, the public versus the private etc.' (Hindess 2005, 394).

Three spirals of power/knowledge

In terms of the workings of power relations and its knowledge effects, the 2007–08 IPY logo describes very well the complexities of the human dimension in polar research. The polar regions are no longer sites of 'bare life', the simple fact of living common to all living beings, but 'bios' which indicated the form or way of living proper to an individual or a group (Agamben 1998). In this sense, the IPY logo is interesting as a symbol of developments in polar regions as human beings cover the globe, instead of the more ordinary symbols of polar regions such as icebergs, polar bears, penguins or indigenous peoples. As Foucault (2007, 41–42) suggests, the environment is one milieu consisting of both natural and human elements: 'the sovereign will be someone who will have to exercise power at that point of connection where nature, in the sense of physical elements, interferes with nature in the sense of the nature of the human species'. These interferences can be seen in terms of three spirals, as a series of sequences of power/knowledge (Michon 2002, 185), spirals of sovereignty and knowledge, discipline and knowledge and finally governmentality and knowledge. These spirals are not static forms of knowledge and power, they are 'matrices of transformations' (quoted in Michon 2002, 174).

The first spiral is about the potential re-emergence of the issue of state sovereignty and knowledge that it requires and produces, particularly in the polar regions, through climate change and its impacts. The importance of territorial control and sovereignty over land, seas, sea routes and natural resources may once again become politically challenged issues. The growth in number of actors in the polar regions, in terms of scientific organizations, indigenous peoples' organizations, environmental organizations and international organizations, should not be seen as a sign of eroding state sovereignty. It is the tactics of power that operates by shifting the work of government from state to non-state agencies (Hindess 2005; Sending and Neumann 2006). As Shadian notes in Chapter 2, when referring to Haas, 'power is now found in many hands and in many places'. The rise of 'global civil society' in its various forms, including scientific cooperation in terms of research projects, research stations and international scientific organizations in the IPY, is part of workings of power, not something outside of it.

The second spiral, discipline and knowledge, is about the role of polar regions in global systems of environmental surveillance. As Jabour and Haward note in their chapter (Chapter 5), there are many similarities from the scientific point of view in these regions. They are important in terms of 'scientific understanding of global processes, and of intrinsic polar processes'. Therefore, while the regions 'are remarkably opposite in many ways, there are fundamental similarities based on scientific utility and the vulnerability of the polar ecosystems to human interference'. The development of international scientific cooperation through networking and institutionalization, the establishment of research stations and the history of the IPY have led to the 'global gaze'. Nilsson points out in Chapter 1 that the 'global gaze from the IGY is now being complemented with bottom-up local gazes that place emphasis on new issues such as the complex interactions of climatic, political, cultural, and social changes. The theme of human dimensions in the International Polar Year 2007–08 can be seen as an expression of this shift'.

Finally, in the third spiral, power is used when defining and framing environmental issues and concerns. This power is often based on the use of scientific knowledge, but the role of science and scientists as the sole authority in these processes of framing is contested by various producers and forms of knowledge. Moreover, through the use of scientific knowledge, environmental problems also become objects for governmental interventions. Because of the neoliberal mode of governmentality, nature is no longer defined and treated as an external, exploitable domain. Through neoliberal processes of capitalization, effected primarily by a shift in representation, previously 'uncapitalized' aspects of nature and society become capitalized (McCarthy and Prudham, 2004). In polar regions, the promises of access to natural resources is a discourse typical to the current neoliberal mode of governmentality in which nature is owned, commodified and commercialized. Previously untapped areas, such as the polar regions, are being opened in the interests of national economies (see Lemke 2000). Neoliberalism, as a mode of governing, is based on 'the concern for the population and its optimization (in terms of wealth, health, happiness, prosperity, efficiency) and the forms of knowledge and technical means appropriate to it' (Dean 1999, 20). Wråkberg points out in Chapter 4 that 'the polar regions as a land of promise for science and a future contributor of wealth to the global economy, is a vision which continues to inspire the post-colonial world'.

The logos of IGY and IPY represent both continuity and change in science and politics in polar regions. These power/knowledge in the polar regions are not marginal for world politics. As Rothwell concludes in Chapter 6, on Antarctic legal and political challenges of the IPY: 'This blending of science, law, politics, and community which is found no where else on the planet has in recent years been reinvigorated by the global focus on climate change and an awareness of the importance of the polar regions as sentinels in climate research'. The developments of polar scientific cooperation and politics of knowledge reflect some important issues in understanding the relationship between knowledge, power and states' roles in world environmental politics. Firstly, the potential for the emergence of territorial claims and close connections to the use of science to make those claims has implications for world politics. Secondly, the concern for the state of the world environment has led to the development of global systems of observing and monitoring, or as I call them in this chapter, systems of surveillance, and the history of setting up this system in polar regions is part of the history of meteorology and climate science. Finally, the use of knowledge in its different forms in defining and framing environmental problems, especially climate change in polar regions, is a challenging area of study and a complex mix of ethical, economic and political considerations.

References

Agamben, G. (1998), *Homo Sacer. Sovereign Power and Bare Life* (Stanford: Stanford University Press).

'Arctic Research Station Belongs in Northwest Passage: Polar Commission' (26 June 2008), CBC News <http.www.cbc.ca/Canada/north/story/2008//06/26/polar research.html>, accessed 18 August 2008.

Burchell, G., Gordon, C. and Miller, P. (eds) (1991), *The Foucault Effect. Studies in Governmentality, with Two Lectures by and an Interview with Michel Foucault* (Chicago: University of Chicago Press).

Canadian High Arctic Research Station (2008), Memo, 4 August 2008.

Chaturvedi, S. (2009), 'Biological Prospecting in the Southern Polar Region: Science–Geopolitics Interface', Chapter 8 in this volume.

Deacon, R. (2002) 'An Analytics of Power Relations: Foucault on the History of Discipline', *History of the Human Sciences* 15:1, 89–117.

Dean, M. (1999), *Governmentality: Power and Rule in Modern Society* (London: Sage).

Foucault, M. (1972), *Archaelogy of Knowledge* (London: Tavistock Publications).

—— (1980), *Power/Knowledge. Selected Interviews and Other Writings 1972–77* (Brighton: Harvester Press).

—— (1991a), 'Questions of method', in Burchell et al (eds).

—— (1991b), 'Governmentality', in Burchell et al. (eds).

—— (2007), *Security, Territory, Population* (Basingstoke: Palgrave/Macmillan).

Hindess, B. (2005), 'Politics as Government: Michel Foucault's Analysis of Political Reason', *Alternatives* 30, 389–413.

Huebert, R. (2009), 'Science, Cooperation and Conflict in the Arctic Region', Chapter 3 in this volume.

Jabour, J. and Haward, M. (2009), 'Antarctic Science, Politics and IPY Legacies', Chapter 5 in this volume.

Lapin Kansa (2008), Venäjä haluaa vetää viralliset rajat napaseudulle, 18 September 2008.

Lemke, T. (2000), 'Foucault, Governmentality and Critique', <http://www.thomaslemkeweb.de/publikationen/Foucault,%20Governmentality,%20and%20Critique%20IV-2.pdf>, accessed 13 June 2008.

McCarthy, J. and Prudham, S. (2004), 'Neoliberal nature and the nature of neoliberalism', *Geoforum* 35:3, 275–83.

Merlingen, M. (2006), 'Foucault and World Politics: Promises and Challenges of Extending Governmentality Theory to the European and Beyond', *Millennium* 35, 181–96.

Michon, P. (2002), 'Strata, Blocks, Pieces, Spirals, Elastics and Verticals: Six Figures of Time in Michel Foucault', *Time & Society* 11:2/3, 163–92.

Nilsson, A.E. (2009), 'A Changing Arctic Climate: More than Just Weather', Chapter 1 in this volume.

Rose, N. (2001), 'The Politics of Life Itself', *Theory Culture Society* 18:6,1–30.

Rothwell, D.R. (2009), 'The IPY and the Antarctic Treaty System: Reflections 50 Years Later', Chapter 6 in this volume.

'Russia Plants Flag Staking Claim to Arctic Region' (2 August 2007), CBC News, <http://www.cbc.ca/world/story/2007/08/02/russia-arctic.html>, accessed 16 June 2008.

Sending, O.J. and Neuman, I.B. (2006), 'Governance to Governmentality: Analyzing NGOs, States and Power', *International Studies Quarterly* 50, 651–72.

Shadian, J. (2009), 'Revisiting Politics and Science in the Poles: IPY and the Governance of Science in Post-Westphalia', Chapter 2 in this volume.

Singer, B.C.J. and Weir, L. (2008), 'Sovereignty, Governance and the Political: The Problematic of Foucault', *Thesis Eleven* 94, 49–71.

Woppke, C.L. (2009), 'The Formation and Context of the Chilean Antarctic Mentality from the Colonial Era through the IGY', Chapter 7 in this volume.

Wråkberg, U. (2009), 'IPY Field Stations: Functions and Meanings', Chapter 4 in this volume.

Index

Note: numbers in brackets preceded by *n* refer to footnotes.

Printed and bound by CPI Group (UK) Ltd, Croydon, CR0 4YY

24/10/2024

01778281-0003